Lecture Notes in Intelligent Transportation and Infrastructure

Series Editor

Janusz Kacprzyk, Systems Research Institute, Polish Academy of Sciences, Warsaw, Poland

The series "Lecture Notes in Intelligent Transportation and Infrastructure" (LNITI) publishes new developments and advances in the various areas of intelligent transportation and infrastructure. The intent is to cover the theory, applications, and perspectives on the state-of-the-art and future developments relevant to topics such as intelligent transportation systems, smart mobility, urban logistics, smart grids, critical infrastructure, smart architecture, smart citizens, intelligent governance, smart architecture and construction design, as well as green and sustainable urban structures. The series contains monographs, conference proceedings, edited volumes, lecture notes and textbooks. Of particular value to both the contributors and the readership are the short publication timeframe and the world-wide distribution, which enable wide and rapid dissemination of high-quality research output.

More information about this series at https://link.springer.com/bookseries/15991

Alexander Schirrer · Alexander L. Gratzer ·
Sebastian Thormann · Stefan Jakubek ·
Matthias Neubauer · Wolfgang Schildorfer
Editors

Energy-Efficient and Semi-automated Truck Platooning

Research and Evaluation

 Springer

Editors
Alexander Schirrer
Institute of Mechanics and Mechatronics
TU Wien
Vienna, Austria

Alexander L. Gratzer
Institute of Mechanics and Mechatronics
TU Wien
Vienna, Austria

Sebastian Thormann
Institute of Mechanics and Mechatronics
TU Wien
Vienna, Austria

Stefan Jakubek
Institute of Mechanics and Mechatronics
TU Wien
Vienna, Austria

Matthias Neubauer
Department of Logistics
University of Applied Sciences
Upper Austria
Steyr, Austria

Wolfgang Schildorfer
Department of Logistics
University of Applied Sciences
Upper Austria
Steyr, Austria

MATLAB and Simulink are registered trademarks of The MathWorks, Inc. See https://www. mathworks.com/trademarks for a list of additional trademarks.

ISSN 2523-3440   ISSN 2523-3459 (electronic)
Lecture Notes in Intelligent Transportation and Infrastructure
ISBN 978-3-030-88684-4   ISBN 978-3-030-88682-0 (eBook)
https://doi.org/10.1007/978-3-030-88682-0

This Springer imprint is published by the registered company Springer Nature Switzerland AG
The registered company address is: Gewerbestrasse 11, 6330 Cham, Switzerland

Foreword by Richard Bishop

Introduction

Platooning is a means for vehicles to safely follow one another in close formation to improve fuel economy and/or improve inter-vehicle efficiency. It relies on "connected braking" between vehicles enabled by direct vehicle-to-vehicle communications. Truck platooning was first implemented and trialled in the late 1990s under the European CHAUFFEUR project. While a key topic of research during the following decade, commercial development of truck platooning burgeoned with the advent of self-driving start-ups in the early 2010s.

First-generation truck platooning has been making steady progress for some time and is now poised for commercial launch. In this mode of operations, the lead driver drives normally, supported by high-function active safety systems. The follow driver steers, monitors the road, and responds as needed to other traffic (for instance, making lane changes to allow traffic to merge onto the highway). Both drivers are responsible for the safe operation of their vehicles. Because automation is applied to longitudinal control only, this is an SAE J3016 Level 1 system.

In second-generation "automated following" platooning systems, the follow truck is driverless, operating at SAE J3016 Level 4, while the lead truck is driven normally. Compared to "solo driverless" systems being developed by many start-ups, the operational design domain of the follow truck is tightly defined, only encompassing the lead truck and any activity in the inter-vehicle gap. With automated follower platooning, fuel economy benefits combine with enhanced driver productivity in the lead truck and labour savings in the follower truck to greatly enhance return on investment.

In Europe, the public and EU member states place a premium on sustainability. Platooning contributes strongly here by reducing energy use and emissions. Even in a zero-emission future with electric trucks, platooning still helps by extending range: less wind resistance translates to more miles on a single battery charge.

Development and Evaluation of Platooning

By now well over 100,000 km of on-road testing of first-generation platooning has occurred across public evaluation projects and private testing by vehicle manufacturers and start-ups. Results from published studies indicate that no safety incidents occurred and fuel economy improvements were confirmed. These studies also found that drivers were highly favourable towards operating trucks in platoon formation.

We can expect to see increased activity in second-generation platooning in the early part of this decade. Automated following is viewed by many as a higher productivity "sweet spot" for platooning. Significant commercialisation activity comes from start-ups developing products intended initially for the US market. These companies see automated following as an accelerated path to initial commercial launch of automated driving, since the human driver in the lead truck can handle a wide range of complex operating scenarios that might challenge a driverless truck operating independently. For instance, the lead truck driver can directly interact with law enforcement and first responders if needed.

Current commercial platooning systems in the USA are designed to operate only on divided highways because that is where the largest number of potential users is operating. Other use cases are now emerging.

The US Army's Next-Generation Combat Vehicle will replace current armoured vehicles with robotic and "optionally manned" vehicles. They aim to use platooning to lessen labour needs for trucked freight, plus reduce exposure of soldiers to hostile action in unsecured corridors. Their "leader–follower" approach might open up gaps between trucks in hostile areas, while tightening the gap in non-hostile areas to gain fuel savings. The army has evaluated an initial fleet of 100 such vehicles and is now preparing to procure thousands more.

The forest industry suffers from severe driver shortages in moving cut timber to mills. Industry players in this domain seek to adapt automated follower platooning to timber transport on back-country roads. This presents extensive engineering challenges, but the urgent business needs are driving a new development programme in Canada which is expected to launch this year.

Deployment of Highway Platooning

Truck speeds, lanes, operations, and fleet sizes differ markedly between Europe and the USA. In Europe, trucks are speed limited to 80 km/h. By contrast, for the vast majority of highway miles in the USA truckers can travel at the same speeds as passenger cars. Therefore, trucks in the US are typically running at higher speeds by 20% or more compared to European operations. Because aerodynamic drag on trucks goes up with speed, the higher the speed the greater the platooning fuel economy benefit. And, because trucks must travel in the rightmost lane in Europe, this limitation in manoeuvring further limits fuel savings potential there. This restriction does not exist in the USA.

In the USA, the top ten fleets by size comprise about 160,000 trucks. The top two fleets, FedEx and UPS, together have over 50,000 trucks. In contrast, the European trucking industry is highly fragmented, and most fleets are relatively small, with the largest trucking companies having about 7000 trucks. Therefore, European fleets may not have large numbers of a single brand of truck, creating the need for brand inter-operability of platooning. This process is well underway, as truck manufacturers have worked together through the Swedish Sweden4Platooning and EU ENSEMBLE projects to develop interoperable data protocols and demonstrate that trucks from different manufacturers can reliably exchange data in two- and three-truck platooning configurations.

Conclusion

Commercially focused platooning development is happening in many forms in many places. Fuel economy improvements from first-generation platooning are a definitive benefit in the USA and countries with similarly structured trucking operations, such as Australia. Globally, the key motivator in the long term is the driver shortage, which is addressed by second-generation platooning. Road capacity, safety improvements and vehicle efficiency are in play as well. Testing and evaluation continue in Europe, where the market is in a formative stage and the preferred platooning configurations have yet to emerge.

We are only a few years away from the introduction of both automated following platooning and highway solo driverless trucking in some form. Initially, solo driverless trucking will be highly constrained, initially operating in low numbers in "good weather" regions. Volumes will rise and steadily expand across more territory as the technology develops to handle challenging weather. By contrast, companies bringing automated following platooning to the market expect to quickly scale across a much broader geography due to having a human in the driving loop.

Given the commercially focused system development and high-level findings so far, this book makes a significant contribution in looking more deeply at potential issues and engineering trade-offs. Important topics addressed include examination of highway throughput, business models, vehicle dynamics and control, digital road infrastructure and interactions between safety, efficiency, comfort and traffic.

In contrast to the US approach which is bottoms-up and rooted firmly in the private sector, the introduction of new technology to European roads is highly collaborative with the public sector and private sector working together, supported by research institutes. Much work has been done, and this book adds extensively to the knowledge base as the way is prepared for beneficial deployment of vehicle platooning.

Highland, MD, USA   Richard Bishop
February 2021   Principal, Bishop Consulting

Foreword by Michael Nikowitz

Transportation is crucial to society and economy, and especially, the road transport sector is changing rapidly. Road freight transportation plays a major role in Austria because its associated demand is expected to increase in the coming years. Hence, the freight transport industry faces big challenges due to steadily increasing energy consumption, mostly based on fossil fuels. Austria claims to be climate neutral by 2040. Therefore, the environmental impacts of all modes of transport need to be reduced urgently by using intelligent mobility systems and innovative technologies.

Automated and connected vehicles promise a solution to meet overall societal goals and mostly climate neutrality. To enable a safe and sustainable introduction of automated mobility, a holistic view on transformative technologies is necessary as there are great expectations and equally much uncertainty. Automated truck platoons can reduce the gap between vehicles on the road and can be seen as a combination of vehicles forming a road train by travelling closely to each other. It is of great benefit to improve fuel economy by reducing fuel consumption during travelling. By focusing on alternative drive trains at the same time, trucks—driven by hydrogen or fully electric—further promise a carbon neutral manner of road transportation. Furthermore, the combination of vehicle and infrastructure has the potential not only to optimise traffic management but also to reduce congestions and dangerous situations. Hence, automated truck platooning offers the potential to increase overall traffic safety and to support the driver in monotonous or critical situation.

However, the overall transformation towards a sustainable, automated and connected mobility is currently facing new challenges. In recent years, the development of automated and highly automated vehicles made huge progress and enabled further fields of applications. Research and development activities with respect to automated truck platoons have turned into a popular topic in recent years as it promises benefits to achieve the societal goals. Presented results from international studies show that a significant fuel reduction potential exists for platoons. Based on the expected advantages and several international studies, truck platoons have attracted the interest of the Austrian Federal Ministry for Climate Action, Environment, Energy, Mobility, Innovation and Technology (BMK). As a result, the flagship project "Connecting Austria" was funded in the frame of the programme "Mobility of the future". With this initiative, the Federal Ministry was focusing on R&I initiatives in respect to connecting of energy-efficient trucks dealing with all aspects from the motorway to the city.

Topics that have to be addressed include determining responsibilities and regulations for a fully integrated sustainable, safe and efficient mobility system as well as highlighting main questions in respect to crossing intersections, entering or leaving highways or approaching hazardous situations. By having a wider perspective on technologies, automation can be assessed on how it can support core targets like the decarbonisation of the transport system, the Vision Zero regarding traffic crashes as well as the effective use of infrastructure by managing mixed fleets. The interdisciplinary project consortium includes scientific institutes, automotive industries and suppliers, infrastructure operators on a national but also international level. Hence, the consortium promises to support the addressed content to set the next steps towards automated driving. Based on topological or weather-related restrictions in Austria, the projects pledge to deliver feedback for the implementation of truck platoons under critical situations and are thus of highest interest for a harmonised European approach and for the automotive supplier industry in general. Results obtained in this project determined criteria under which conditions the operation of automated truck platoons is possible and points out further actions, which should be taken to guarantee a safe and sustainable use of this technology. It also highlights measures that are a prerequisite for a successful integration of automation in our transport system and demonstrates that present systems can be enhanced significantly through the introduction of automated truck platoons. Based on this result, the Federal Ministry, the Austrian automotive supplying industry and the scientific sector are well prepared and capable of dealing with the next steps for development and implementation within the existing ecosystem.

Vienna, Austria   Michael Nikowitz
February 2021   Coordinator Automated Mobility
                                             Federal Ministry for Climate Action,
                                                 Environment, Energy, Mobility
                                                   Innovation and Technology

Preface

Cooperative, connected automated mobility (CCAM) provides many opportunities to improve traffic system efficiency and traffic safety, as well as to reduce emissions. However, it also faces many challenges and open questions in technological aspects, legal matters, user acceptance or logistics' requirements. This was basically the reason for the Austrian Federal Ministry of Transport, Innovation and Technology to open a tender with regard to automated driving and logistics in 2017. The Austrian flagship project "Connecting Austria" was selected to be funded within this call. The project targeted mainly the truck platooning use case and took a multidisciplinary research approach to address these issues broadly. This book highlights a selection of obtained results. The book targets on the one hand readers with an engineering background interested in current and state-of-the-art methods and techniques to realise semi-automated driving and embeds such vehicles into a smart traffic ecosystem. On the other hand, it is intended to be accessible to a more general audience clearly addressing technological capabilities and limitations and their impact on safety, societal or economic aspects of automated driving and truck platooning in particular. The project's interdisciplinary research effort considered aspects in engineering, road vehicle technology, road infrastructure technology and traffic management and optimisation, but also traffic safety, user acceptance issues including psychological open questions and relevant economic aspects. This unique multidisciplinary research approach and an outlook on open questions and next steps with regard to CCAM make this book a unique contribution to the body of existing literature.

Synopsis

Truck platooning may represent a core element towards the sustainable and safe transport of goods. This book provides insights into the design and deployment of energy-efficient, safe and semi-automated truck platoons with a special focus on connected cooperative automated mobility (CCAM). The book contains the following three main parts. *Part I—Contextualising Truck Platooning, Part II—Assessment Methodologies and Their Application* and *Part III—Towards Cooperative Truck Platooning Deployment.*

Part I summarises contributions of truck platooning projects worldwide and relevant deployment requirements from the different perspectives like energy efficiency, traffic safety or cooperative intelligent transport systems (C-ITS). Furthermore, an overview related to the Austrian research project "Connecting Austria" provided and challenges when it comes to potential benefits of CCAM as well as possible trade-offs are discussed.

After setting the scene for truck platooning in Part I, truck platooning assessment methodologies and their application in the Connecting Austria project are detailed in Part II. Due to the heterogeneous and multidisciplinary research endeavour, research results relevant for different research domains are summarised in this part. The results cover investigations of aerodynamic drag effects due to reduced inter-vehicle distances and their validation on test tracks. Furthermore, study results related to comprehensive investigations of truck platoon dynamics and traffic effects are presented in Part II. Thereby, also the facilitation of truck platooning through intelligent road infrastructures, which are able to monitor traffic and provide relevant information to other road participants, is investigated. Finally, Part II presents an approach to assess truck platoon efficiency and its application within the Connecting Austria project.

Part III outlines paths towards cooperative truck platooning deployment. In doing so, traffic safety-related issues in general and specific issues in Austria are discussed as a main prerequisite for automated driving. Moreover, taking a road operator's view, business model aspects relevant to the C-ITS deployment are assessed to reveal sustainable next deployment steps. The remainder of this part investigates advanced powertrain systems to increase sustainability of truck platoons and the impacts of truck platooning research on the European innovation system. Part III concludes with a general discussion on lessons learnt and open questions regarding CCAM.

Linz, Austria   Wolfgang Schildorfer
February 2021

Acknowledgements

The authors and editors are grateful for the dedication, enthusiasm and continuing support of the Connecting Austria consortium and the involved individual work groups. The results presented in this book were developed within the Austrian flagship project on connected and automated mobility "Connecting Austria". Special thanks to the Austrian Federal Ministry for Climate Action, Environment, Energy, Mobility, Innovation and Technology who financially supported our project. Furthermore, we want to thank the Austrian Research Promotion Agency (FFG) for the professional administration of the project. Connecting Austria was funded under the FFG grant no. 865122.

Contents

Editors and Contributors

About the Editors

Alexander Schirrer has been Postdoctoral Researcher and Teacher of graduate-level lectures in the Institute of Mechanics and Mechatronics, TU Wien, since 2011. His research interests include modelling, simulation, optimisation and control of complex and distributed parameter systems.

Alexander L. Gratzer has been Project Assistant in the Institute of Mechanics and Mechatronics, TU Wien, since 2019, and currently works towards the Ph.D. degree. He studied mechanical engineering and is involved in international research projects and teaching of graduate-level lectures. His research interests include simulation, optimisation and control of complex industrial systems.

Sebastian Thormann has been Project Assistant in the Automation and Control Institute, TU Wien, since 2021, and currently works towards the Ph.D. degree. He studied mechanical engineering and was a student member of the project team in the Institute of Mechanics and Mechatronics, TU Wien, from 2017 to 2020. His research interests include modelling, simulation, optimisation and control of complex and distributed parameter systems.

Stefan Jakubek is Professor and Head of the Institute of Mechanics and Mechatronics, TU Wien. From 2007 to 2009, he was Head of Development for Hybrid Powertrain Calibration and Battery Testing Technology in AVL List GmbH, Graz, Austria. His research interests include fault diagnosis and system identification.

Matthias Neubauer is Professor at the University of Applied Sciences Upper Austria for logistics information systems. His research interests cover human–computer interaction, intelligent transportation systems as well as cooperative, connected and automated mobility. He is involved in international and national research projects and teaches in the master's programme digital transport and logistics management classes such as process management, distributed logistics systems or geoinformation systems.

Wolfgang Schildorfer is a person who likes the road he still walks to find new chances. Since October 2018, he has been Professor for Transport Logistics and Mobility at the University of Applied Sciences Upper Austria. His research focus is on innovation, business models and evaluation in transport logistics, smart hyper-connected logistics systems (urban) mobility, sustainable transport systems and new technology markets (C-ITS, CCAM, automated driving, truck platooning).

Contributors

Walter Aigner High Tech Marketing, Vienna, Austria

Hatun Atasayar Austrian Road Safety Board (KFV), Vienna, Austria

Philipp Blass Austrian Road Safety Board (KFV), Vienna, Austria

Patrick Brandtner Department of Logistics, University of Applied Sciences Upper Austria, Steyr, Austria

Ilja Bäumler Universität Bremen, Bremen, Germany

Almir Cajic Virtual Vehicle Research GmbH, Graz, Austria

José Carmona Andata Entwicklungstechnologie GmbH, Hallein, Austria; Andata Artificial Intelligence Labs, Hallein, Austria

Christian Doppler Virtual Vehicle Research GmbH, Graz, Austria

Bernhard Fischbacher Virtual Vehicle Research GmbH, Graz, Austria

Alexander L. Gratzer Research Unit of Control and Process Automation, Institute of Mechanics & Mechatronics, TU Wien, Vienna, Austria

David Hildenbrandt Andata Entwicklungstechnologie GmbH, Hallein, Austria; Andata Artificial Intelligence Labs, Hallein, Austria

Thomas Hoch Software Competence Center Hagenberg GmbH, Hagenberg, Austria

Florian Hofbauer Department of Logistics, University of Applied Sciences Upper Austria, Steyr, Austria

Christoph Irrenfried Graz University of Technology, Institute of Fluid Mechanics and Heat Transfer, Graz, Austria

Stefan Jakubek Research Unit of Control and Process Automation, Institute of Mechanics & Mechatronics, TU Wien, Vienna, Austria

Susanne Kaiser Austrian Road Safety Board (KFV), Vienna, Austria

Theodorich Kopetzky Software Competence Center Hagenberg GmbH, Hagenberg, Austria

Alexander Kospach Virtual Vehicle Research GmbH, Graz, Austria

Herbert Kotzab Chair of Logistics Management, Universität Bremen, Bremen, Germany

Andreas Kuhn Andata Entwicklungstechnologie GmbH, Hallein, Austria; Andata Artificial Intelligence Labs, Hallein, Austria

Bernhard Lechner Virtual Vehicle Research GmbH, Graz, Austria

Andrea Massimiani Department of Logistics, University of Applied Sciences Upper Austria, Steyr, Austria

Alexander Mladek Virtual Vehicle Research GmbH, Graz, Austria

Matthias Neubauer Department of Logistics, University of Applied Sciences Upper Austria, Steyr, Austria

Thomas Novak SWARCO FUTURIT Verkehrssignalsysteme GmbH, Perchtoldsdorf, Austria

Michael Nöst IESTA—Institute for Advanced Energy Systems and Transport Applications, Graz, Austria

Peter Sammer Virtual Vehicle Research GmbH, Graz, Austria

Oliver Schauer Department of Logistics, University of Applied Sciences Upper Austria, Steyr, Austria

Wolfgang Schildorfer Department of Logistics, University of Applied Sciences Upper Austria, Steyr, Austria

Alexander Schirrer Research Unit of Control and Process Automation, Institute of Mechanics & Mechatronics, TU Wien, Vienna, Austria

Elvira Thonhofer Andata Entwicklungstechnologie GmbH, Hallein, Austria; Andata Artificial Intelligence Labs, Hallein, Austria

Sebastian Thormann Research Unit of Control and Process Automation, Institute of Mechanics & Mechatronics, TU Wien, Vienna, Austria

Gerold Wagner Department of Logistics, University of Applied Sciences Upper Austria, Steyr, Austria

Erwin Wannenmacher Austrian Road Safety Board (KFV), Vienna, Austria

Martin Winkelbauer Austrian Road Safety Board (KFV), Vienna, Austria

Christoph Zitz Virtual Vehicle Research GmbH, Graz, Austria

Michael Zotz Virtual Vehicle Research GmbH, Graz, Austria

Part I
Contextualising Truck Platooning

Part I summarises contributions of truck platooning projects worldwide and relevant deployment requirements from the different perspectives like energy efficiency, traffic safety or cooperative intelligent transport systems (C-ITS). Furthermore, an overview related to the Austrian research project "Connecting Austria" provided and challenges when it comes to potential benefits of CCAM as well as possible trade-offs are discussed.

Chapter 1
Connecting Austria Project Outline

Walter Aigner, Andreas Kuhn, Thomas Novak, and Wolfgang Schildorfer

Abstract In 2017, the core team of the Connecting Austria project faced the challenge of leveraging previous research results on cooperative intelligent transport systems (C-ITS) into the logistics domain—namely into the domain of truck platooning. Quite a lot of ideas and topics were evaluated, potential research partners explored, and funding opportunities for a cooperative research project were assessed. The window-of-opportunity opened in 2017 when the Federal Ministry on Transport, Innovation and Technology started a tender for a flagship research project on automated driving in different domains. This was the start of "Connecting Austria". The following paragraphs outline the project in a nutshell, the project objectives, technology domains targeted and the planned test procedure, use cases and finally sketch the challenges and international uniqueness of the Connecting Austria project.

Keywords Platooning use cases · C-ITS (cooperative intelligent transport systems)

1.1 Connecting Austria in a Nutshell

The flagship project Connecting Austria brings technology leaders and end-users together to demonstrate and evaluate four specific use cases for semi-automated and energy-efficient truck platoons. Key objectives is the evidence-based evaluation of energy-efficient truck platoons as a prerequisite for competitive strength of Aus-

W. Aigner
HiTec Marketing, Lothringerstraße 14/6 / A -1030, Vienna, Austria

A. Kuhn
Andata Entwicklungstechnologie GmbH, Hallburgstraße 5, 5400 Hallein, Austria

T. Novak
SWARCO FUTURIT Verkehrssignalsysteme GmbH, Mühlgasse 86, 2380 Perchtoldsdorf, Austria

W. Schildorfer (✉)
Department of Logistics, University of Applied Sciences Upper Austria, Steyr, Austria
e-mail: wolfgang.schildorfer@fh-steyr.at

© The Author(s) 2022
A. Schirrer et al. (eds.), *Energy-Efficient and Semi-automated Truck Platooning*,
Lecture Notes in Intelligent Transportation and Infrastructure,
https://doi.org/10.1007/978-3-030-88682-0_1

trian industries such as logistics, telematics and infrastructure providers, automotive suppliers, as well as vehicle development and cooperative research.

The national flagship project's unique contribution is its specific focus on infrastructure issues and on parametrised traffic perspectives when evaluating energy-efficient and semi-autonomous truck platoons. This particularly includes platoons at intersections before entering motorways and after leaving motorways.

Connecting Austria defines a truck platoon as follows: The platoon consists of two to maximum three trucks, the automation level is SAE-L1 or SAE-L2, every truck is led by a truck driver with always keeping his hands on the wheel, and the distance between the trucks for several impact analysis is about 15 m.

Connecting Austria leverages Austrian strategic strengths as pioneer in C-ITS infrastructure and continues international success stories such as ECo-AT (Austrian part of the C-ITS Corridor), coordination activities in C-Roads, as well as the pioneering role in vehicle expertise (European Green Car Initiative, electric power train in trucks). Connecting Austria is going to share its findings with European and international "Cooperative Connected and Automated" projects and initiatives (EU, DG MOVE, DG CONNECT).

Project data:

- Duration: 36 months.
- Project start: 01/01/2018.
- Project budget: 4.3 MEuro.
- Project funding (bmvit): 2.5 MEuro.
- Web: www.connecting-austria.at.
- Project Leader: Dr. Wolfgang Schildorfer.

1.2 Connecting Austria's Objectives

Key objective of the project was defined as the evidence-based evaluation of energy-efficient truck platoons as a prerequisite for competitive strength of Austrian industries such as logistics, telematics and infrastructure providers, automotive suppliers, as well as vehicle development and cooperative research.

Further objectives were defined in the project proposal as follows:

- Regarding technological targets, the feasibility and limits of energy-efficient, semi-automated and connected truck platoons in ASFINAG's road network between Hallein via Linz/Pasching to Vienna and selected streets with traffic lights and intersections will be validated.
- The commercial freight traffic target is the validation of energy savings of 15% reported by truck manufacturers in case of changing from Adaptive Cruise Control – ACC (distance of about 50 m) to semi-automated and energy-efficient truck platoons (distance of about 15 m) and the validation of the total costs of the change.

- Regarding climate and sustainability targets, Connecting Austria aims at delivering evidence-based simulations and projections of the effects of energy-efficient and semi-automated platoons on the whole traffic infrastructure including all participants on the road and including future electric power train developments.
- The economic target is defined to give a quality-proven answer to the key question for a national flagship project: How can Connecting Austria strengthen the competitiveness of one or more commercial sectors (shippers, logistics, telematics, automotive suppliers, vehicle development and cooperation research) and improve the competence of research-intensive organisations, including the horizontal and vertical integration of the value chain?
- The organisational target is defined to identify necessary measures for the implementation of truck platoons in Austria.
- Concerning road safety targets, methods and tools that evaluate road safety of semi-automated truck platoons have to be developed. Is it possible to improve road safety with cooperative control strategies?
- With regard to traffic optimisation, it will be evaluated if cooperative vehicle and traffic control strategies have positive effects on the traffic, e.g. improved traffic efficiency, prevention of traffic jams. Connecting Austria will answer this question and provide a decision base for traffic planning.
- An additional target is to spread project results out into the public. Questions on innovation politics, ethics, governance, road safety and IT security will be discussed in the context of automated driving.

1.3 Technology Domains of Connecting Austria and the Planned Testing Procedure

The project's focus in the proposal was on three technology domains:

- Sensor technology.
- Control strategies for vehicles and infrastructure.
- Data exchange technology.

Sensor Technology is the key component regarding automated driving and truck platoons. In the project, the infrastructure-based sensor technology and data exchange are in the focus of research. A mix of sensor types should detect traffic participants that are not connected and in danger (e.g. pedestrians and motorcyclists) predict their behaviour. This functionality is crucial to safely drive a truck platoon across an intersection, ensuring the safety of all involved traffic participants.

Control strategies for vehicles and infrastructure are necessary to form a truck platoon, to maintain it and to go back to a regular transport mode. From an infrastructure perspective, criteria such as traffic situation, city, countryside or motorway are the basis for the decision if a platoon can be formed or has to be dissolved in a specific area. In the vehicle, sensor data coming from the infrastructure will be connected with data coming from the vehicle to evaluate the local circumstances in short feedback circles. Consequently, it can be determined if the platoon will be set up autonomously, maintained, or dissolved.

Data exchange is the third essential element for the infrastructure- and vehicle-based management of truck platoons. Based on ITS-G5 technology and the Europe-wide harmonised message set, data elements (e.g. platoon permission from a to b) are exchanged between vehicles (V2V) and infrastructure (V2I) (C-ITS Day2 Use Case—Cooperative Platooning).

1.4 Connecting Austria Use Cases

Four use cases were defined for the core research work in Connecting Austria. Those use cases were the guiding principle for the whole project: (1) Trucks entering the motorway and forming a platoon, (2) truck platoon approaching a hazardous location, (3) truck platoon leaving the motorway and (4) truck platoon crossing an intersection. The following paragraphs outline the four use cases including a basic scenario description:

1.4.1 Use Case 1: Trucks Entering the Motorway

Scenario: Three trucks drive form the shipping point to the motorway. The trucks are informed by a road side unit, if it is suitable and permitted to form a platoon in this specific motorway section. Criteria for such adaptive permissions are, for example, traffic situation or the environmental status. The trucks form a platoon and transmit the information about the platoon's status via on-board unit, compare Fig. 1.1.

1.4.2 Use Case 2: Truck Platoon Approaching a Hazardous Location

Scenario: Three trucks have formed a platoon and approach a hazardous location. The road-side unit requests them to dissolve the platoon. The hazardous location (e.g. road works) does not allow any platoon for a certain period of time. The trucks dissolve the platoon and transmit the information about the platoon's status, compare Fig. 1.2.

1.4.3 Use Case 3: Truck Platoon Leaving the Motorway

Scenario: A platoon consisting of three trucks approaches the exit of the motorway. One truck intends to leave the motorway, and the others continue driving on the motorway. The trucks get the information via the road-side unit to dissolve the platoon when leaving. Consequently, they dissolve the platoon and one truck leaves the motorway while the other two trucks remain on the motorway, compare Fig. 1.3.

1.4.4 Use Case 4: Truck Platoon Crossing an Intersection

Scenario: A platoon consisting of three trucks drives on the road approaching an intersection with traffic lights for pedestrians, prioritised public transport and other traffic participants. The platoon transmits information about its status and its intention, respectively, via an on-board unit. Based on criteria like daytime, presence of pedestrians, prioritisation of public transport or traffic situation, the road-side unit transmits information to the platoon on how to cross the intersection under best possible conditions, either by dissolving, stretching or maintaining the platoon. Compare Fig. 1.4.

Fig. 1.1 UC1: Trucks entering the motorway. © 2018 Swarco Futurit, reproduced with permission

Fig. 1.2 UC2: Truck platoon approaching a hazardous location. © 2018 Swarco Futurit, reproduced with permission

Fig. 1.3 UC3: Truck platoon leaving the motorway. © 2018 Swarco Futurit, reproduced with permission

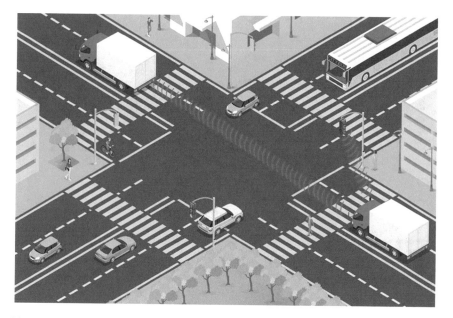

Fig. 1.4 UC4: Truck platoon crossing an intersection. © 2018 Swarco Futurit, reproduced with permission

1.5 Challenges, International Uniqueness and Discussion

When starting the project in 2017, the main challenge was to answer the following questions. How may Austria, as a small transit country with ambitious climate goals, proactively shape the next technology for energy-efficient and automated freight transport in line with social challenges such as Vision Zero (traffic safety), electro mobility? Furthermore, how may Austria handle the transition to a heterogeneous vehicle population, as well as resolve traffic efficiency challenges, i.e. improved use of existing interurban road infrastructure and increasing goods transport? In addition, the project team asked the question how Austria may strengthen the sustainability skills of a traditionally strong telematics industry (e.g. C-ITS provider or even the Austrian road operator ASFINAG) and the logistics industry.

The above-mentioned challenges were indented to be managed by providing evidence-based answers within the project using a stepwise testing approach, compare Fig. 1.5.

Simulation: Components, subsystems as well as their performance in the entire system including vehicles, traffic and infrastructure factors are tested by complex numeric simulations.

Tests in a closed environment: Vehicle- and infrastructure-based subsystems and their integration are tested with project and cooperation partners in a closed environment.

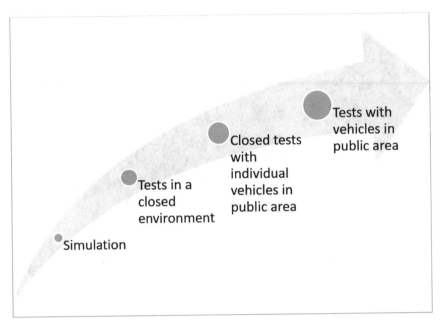

Simulation

Tests in a closed environment

Closed tests with individual vehicles in public area

Tests with vehicles in public area

Fig. 1.5 Connecting Austria—planned testing procedure

Closed tests with individual vehicles in public area: Vehicle- and infrastructure-based subsystems and their integration are tested with project and cooperation partners on public roads. The initial plan in Connecting Austria comprised test drives of the fleet operator TRANSDANUBIA carried out without freight due to safety reasons. This is a common approach, which was also applied by the comparable EDDI project in Germany. A prerequisite for tests with freight is the successful accomplishment of test drives without freight.

Tests with vehicles in public areas: Vehicle- and infrastructure-based subsystems and their integration are tested within public testing areas on public roads. The initial plan in Connecting Austria was to integrate truck platoons in real logistics operations together with OEMs such as VOLVO, DAF, MAN, IVECO and DAIMLER. Unfortunately, the legal prerequisites for truck platooning in Austria could not be met within the project duration. However, important preparatory work for legal regulations could be developed within the project as starting point for follow-up projects.

When setting up the project in 2017, it was internationally unique in the special consideration of the intelligent infrastructure and traffic perspective and in the special consideration of investigating an urban truck platooning use case with traffic-light-controlled intersections before and after motorway entrances. The three main target groups of the project were: (1) road operators/infrastructure providers, (2) logistics operators, and (3) C-ITS industry. Especially for those target groups and policymaker, one common topic was the guiding principle—"How can safe truck platooning reduce

CO_2-emissions and how can this help to strengthen the stakeholders' role in their market or political environment?".

When setting up the project in 2017, it was internationally unique in the special consideration of the intelligent infrastructure and traffic perspective and in the special consideration of investigating an urban truck platooning use case with traffic-light-controlled intersections before and after motorway entrances. The three main target groups of the project were: (1) road operators/infrastructure providers, (2) logistics operators, and (3) C-ITS industry. Especially for those target groups and policymaker, one common topic was the guiding principle—"How can safe truck platooning reduce CO_2-emissions and how can this help to strengthen the stakeholders' role in their market or political environment?".

Within the CCAM domain, the infrastructure role for supporting truck platoons was not clear and validated at project start. Connecting Austria investigated this important infrastructure provider role with special focus on C-ITS (ITS-G5) technology to support truck platooning on the Austrian motorways. A main question to be answered was "How can potential benefits with regard to cost savings and CO_2-reduction be positively affected by a dynamic traffic management of a road operator?".

From a logistics domain point of view, the need for innovation arose from demand to meet CO_2 targets and improve efficiency. Connecting Austria provided insights on potential efficiency gains that allow to project efficiency gains for logistics operators.

Finally, the third group—the C-ITS industry—has faced some challenges with regard to CCAM and CO_2-emission reduction. It had not been clear how C-ITS provider can proactively support traffic management in urban areas as well as on motorways to decrease CO_2-emissions with dynamic traffic management tools. Especially for the truck platooning use case, the need for assessing the direct impact of C-ITS-based dynamic traffic management was the motivation for this research project.

This book provides some answers to those above-mentioned challenges as well as interesting truck platooning-related research results. Although, open public testing with SAE-L1 or SAE-L2 truck platoons from different truck OEMs was not possible within the Connecting Austria project, very interesting research results with regard to the impact of C-ITS dynamic traffic management on CO_2-emissions, truck platooning control concepts or traffic safety requirements could be achieved.

Walter Aigner has been a kind of boundary-spanning individual [between pioneering users, public administration, research and various industries] since the early 1990s. As managing director at HiTec he prepared several national and European innovation and technology programmes and serves as an independent expert on evaluation and impact assessment. His focus is on key individuals in the European innovation system and how they nourish our commitment to contributing to a more nuanced answer for Europe's share in a global challenge to effectively cooperate with the US and Asia in a highly competitive environment of innovation, digitalisation and automation.

Andreas Kuhn studied Technical Mathematics and Mechanical Engineering at the Vienna University of Technology. There he also awarded his Ph.D. for the simulation of special satellite dynamics. He now works for more than two decades in several positions and roles in the fields of automotive safety, automated driving and traffic automation with an steady focus on virtual development procedures and the safe application of softcomputing methods.

Thomas Novak holds a Ph.D. in electrical engineering from the TechnicalUniversity of Vienna. Since 2008, he is with the Austrian company SWARCO FUTURIT. Starting as software project manager and later on as innovation manager, he is now working as portfolio manager. The focus is on the deployment and go2market of CCAM solutions in Europe.

Wolfgang Schildorfer is a person who likes the road he still walks to find new chances. Since October 2018, he has been Professor for Transport Logistics & Mobility at the University of Applied Sciences Upper Austria. His research focus is on innovation, business models and evaluation in transport logistics, smart hyperconnected logistics systems, (urban) mobility, sustainable transport systems and new technology markets (C-ITS, CCAM, automated driving, truck platooning).

Chapter 2
Truck Platooning Worldwide

Hatun Atasayar, Philipp Blass, and Susanne Kaiser

Abstract Although early attempts date back several decades, truck platooning initiatives and trials gained significant momentum in the 2010s in the light of the overall wave of vehicle automation. This chapter aims at providing an overview of the most significant platooning endeavours over the past two decades, with focus on European efforts and their thematic priorities and conclusions. Whereas for example relevant Southeast Asian countries tend to be driven by national roadmaps and national funding, many of the European projects and trials are funded by the European Commission's research programmes, bringing together several countries and resulting in more open access information. The presented review demonstrates how the focus of the discussion changed over time, also revealing a shift in the relevant parties and stakeholders: while the initial intention was reduction in fuel costs, the claimed benefits have now diversified and additional interest groups brought to the table environmental, safety, legal, infrastructural and labour regulation arguments for and against platooning.

2.1 Introduction

There are several initiatives, research activities and studies on truck platooning across the globe, driven by the European Union, various governments and the industry. Truck platooning is the linking of two or more trucks in convoys, using automated driving support systems. These vehicles automatically maintain a close distance between each other when they are connected for certain parts of a route. The truck at the head of the platoon acts as the leader, with the vehicles behind reacting and adapting to changes in its movement, e.g. accelerating or braking. Drivers of the trailing vehicles remain in control of their vehicle, so they can leave the platoon and drive independently. [1] Truck platooning has great potential to make road transport safer, cleaner and more efficient in the future [1]. Therefore, truck manufacturers are striving to develop and deploy truck platooning systems. First trials have been

H. Atasayar (✉) · P. Blass · S. Kaiser
Austrian Road Safety Board (KFV), Vienna, Austria
e-mail: hatun.atasayar@kfv.at

© The Author(s) 2022
A. Schirrer et al. (eds.), *Energy-Efficient and Semi-automated Truck Platooning*,
Lecture Notes in Intelligent Transportation and Infrastructure,
https://doi.org/10.1007/978-3-030-88682-0_2

13

carried out by several projects and initiatives. However, further research on truck platooning technology is necessary before truck platoons are permitted to operate widely on public roads. Experience with recent truck platooning technologies under real traffic conditions is a prerequisite to, e.g., be able to assess how other road users react to truck platoons and what the ideal number of vehicles in a platoon is. With each trial under real traffic conditions, new challenges and opportunities can be derived together with relevant stakeholders. This includes not only OEMs, Tier 1 suppliers, software and service providers but also road operators, logistics companies, insurance companies and policy makers.

With the (further) development of driver assistance systems, connectivity, sensor technology and the digitalisation of the traffic system, truck platooning has gained in importance. While truck platooning was initially seen primarily to reduce fuel consumption, the discussion soon shifted to how truck platooning can contribute to a sustainable transport system. Major research projects and trials were conducted in the USA, Europe, Asia and Australia to evaluate the benefits and feasibility of truck platooning. In California, the PATH programme has been conducting trials for years. The European project Promote Chauffeur I was one of the first demonstrations of truck platooning with two trucks using tow-bar technology. In the continuation of the project in Promote Chauffeur II, the feasibility of a fully operable truck platoon with three trucks was demonstrated under real conditions [2]. The German KONVOI project investigated the advantages and operational issues of truck platoons used in mixed traffic scenarios on the motorway [3]. The European SARTRE project demonstrated a platoon consisting of both passenger cars and trucks on the motorway, using a manually operated truck trailed by automated passenger cars [4]. This is only a fraction of trials with such and related topics around the world.

2.2 Opportunities and Challenges of Truck Platooning

2.2.1 Interoperability

In Europe, given the predominance of small fleets with only few trucks of a single brand, interoperability between multi-fleet and multi-brand trucks plays a key role. The project Sweden 4 Platooning, e.g., demonstrated the feasibility of truck platoons with vehicles from different manufacturers in DB Schenker's operations. This was achieved by harmonising the systems of the manufacturers Scania and Volvo. The research project COMPANION (cf. [5]) investigated possibilities for the application of the truck platooning concept for commercial freight transport. For this purpose, a coordination system was developed within the project, which enables the dynamic formation and dissolution of truck platoons. This means that vehicles participating in a platoon do not have to have the same origin and/or final destination but can also run together on parts of their route. Furthermore, it was investigated how truck drivers can be informed when a potential truck platoon can be formed or dissolved. The project

CONCORDA (cf. [6]) also focuses on the interoperability and networking between different systems. A core objective of the project is to improve the interoperability of technologies, services and their implementation in the European Union. The European project ENSEMBLE (cf. [7]) investigates the impact of multi-brand platooning on infrastructure, drivers, traffic safety and traffic flow. The project plans to test multi-brand platooning on test tracks and on public roads across national borders in 2021.

2.2.2   Road Safety and Traffic Efficiency

According to the European Truck Platooning Challenge, human error is responsible for more than 90% of road accidents and the human factor is decisive when it comes to traffic efficiency. Several projects examined the advantages, disadvantages, impact and safety effects of truck platooning in simulation studies, on test tracks as well as in real traffic situations in the last two decades. Subsequently, results from selected projects that evaluated interactions with other road users and driver perception are described.

The focus of the KONVOI project (cf. [3]) was on impact analysis (driver acceptance, traffic flow, environment) and the investigation of legal and economic implications of truck platooning. During the project a platoon of four trucks of the brands MAN and IVECO was formed and tested in real traffic for the first time worldwide in 2009. The participating vehicles were equipped with vehicle-to-vehicle (V2V) and vehicle-to-infrastructure (V2I) communication, a mono-camera as well as a lidar and radar sensor and could thus travel at distances of 10 m between each other. Based on real-life traffic (a total of 3100 km were covered), the project claims to have demonstrated safe traffic flow. However, no details on the results (methods, parameters, design, etc.) are publicly available. The German project EDDI (cf. [8, 9]) was able to analyse the experience of truck drivers and their interaction with other road users as well as psychosocial and neurophysiological effects of truck platooning on truck drivers. The project involved seven months of testing with two linked vehicles on the motorway between Nürnberg and München (approximately 165 km). The results indicate a significant change in the previously partly critical attitude of the drivers. The project was tested for the duration of seven months with two linked vehicles. For the international use, the scientists recommend further investigations with longer platooning phases. In the PLATOON project (cf. [10]), the advantages and disadvantages of truck platooning regarding transport operations were investigated to understand the effects of platooning strategies on transport operations on freeways. To this end, these effects were theoretically analysed to determine the limits of truck platooning and thus to decide when and how the formation of a platoon can improve motorway traffic operation. Traffic performance and safety effects of different platooning strategies and platoon configurations using microscopic simulation have been analysed in the paper "Benefits and risks of truck platooning of freeway operations near entrance ramp" (cf. [11]). The driving behaviour of truck platoons and conventionally controlled vehicles was also modelled for the same paper. The simulation showed that truck platoons may lead to problems at freeway entrances: The greater the traffic volume and the higher the number of truck platoons, the more vehicles may not be able to fit into the flow of traffic within the length of the acceleration lane. These vehicles must either stop at the end of the acceleration lane or continue along the hard shoulder. These manoeuvres are associated with a higher risk of accidents and affect the fluidity of the traffic. Therefore, it is recommended to allow truck platooning on on-ramps only under certain conditions (e.g. only at certain times of day). In case of higher traffic volume—especially at on-ramps—the use of truck platoons is not recommended.

2.2.3 Operation Costs and Fuel Consumption

The use of platooning technology/systems in the truck industry can help vehicle manufacturers and transport service providers to reduce fuel costs. According to Scania, fuel costs account for more than 30% of operating costs in normal European transport operations. Reducing fuel consumption would therefore have a significant impact on total freight costs in the transport industry. For example, tests conducted by Scania in December 2015 showed that a truck platoon can reduce fuel consumption by up to 12%. According to Peloton Technology, reducing drag on two-truck platoons offers fuel savings for both the trailing and the leading truck. However, the reduction in fuel consumption is highly dependent on the speed of the trucks. On European motorways, the speed of trucks is limited to 80 km/h, while in the USA, trucks travel at the same speed as cars for most motorway kilometres. Since air resistance increases with speed, fuel savings in the USA and in countries with similar truck driving patterns, such as Australia, are expected to be greater than in Europe. A detailed comparison of fuel savings reported in diverse truck platooning projects is provided in Chap. 3. Thereby, the comparison considers different following distances within a truck platoon and the speed of a truck platoon.

2.2.4 Reduction of CO_2 Emissions

The International Transport Forum (ITF) estimates that international trade-related freight transport is currently responsible for around 30% of all transport-related CO_2 emissions from fuel combustion and for more than 7% of global emissions. In order to achieve the Parisienne climate targets, OEMs and Tier 1 suppliers are expected to comply with climate protection requirements. The European Parliament is calling for a 30% reduction in CO_2 values for trucks and buses by 2030 (reference year 2019). The reduction of the fuel consumption is directly related to the reduction of CO_2 emissions. As such, truck platooning may represent one means to reduce CO_2

emissions. An assessment of the emission reduction potential for a certain case is detailed in Chap. 12.

2.2.5 Shortage of Professional Drivers

According to the German Federal Motor Transport Authority, about 20% of professional drivers are over 55 years old and about to retire in the years to come. Only 40% of the vacant positions due to retirement can be reoccupied (67,000 professional drivers retire annually and only 27,000 junior drivers follow (cf. [12]). For this reason, a higher degree of automation in the transport sector is considered to solve the shortage of professional drivers. At the same time, the increasing level of automation is expected to rise the attractiveness of the profession, as the latest technology will require drivers with new skills and thus open new areas of responsibility. To solve the driver shortage issue, the Japanese government plans to commercialise unmanned trailing platooning vehicles by 2022.

2.2.6 New Requirements for Vehicles and the Infrastructure

The transition from conventional transport systems towards automated and connected mobility (digitalisation of the traffic system) sets new challenges, which require constant adjustments to both vehicles and infrastructure. On the vehicle side, the installation of advanced automated driving features from different OEMs and vehicle generations will lead to a huge range of different versions of software capability, which must be able to communicate and interact with each other and the infrastructure (V2V, V2I and V2X). The interaction of vehicles and infrastructure is achieved by collecting, processing and intelligently linking data (cf. "Strategy for Automated and Connected Driving" defined by the German government in 2015). The basic prerequisite is the generation of secure and fast data transmission, which results in high functional and qualitative demands on the communication technology and infrastructure. Most of the current road infrastructure (globally) is not sufficiently well equipped and thus restricts the formation of routes for long-distance truck platoons. Furthermore, unsuitable road markings and bridges hamper an optimised route for truck platooning. Therefore, before truck platoons can become a common sight on Europe's road, the road infrastructure needs to be upgraded (cf. [1]).

ASFINAG, the Austrian motorway operator, is equipping the Austrian freeway network with C-ITS starting in November 2020 (cf. [13]). The C-ITS rollout has been coordinated in Europe with operators in 18 member states and automobile manufacturers. The goal is to network directly with vehicles and to be able to support future applications from electromobility to highly automated driving. In 2018, the Korean government set up a 7.7 km long test track on the Yeonju Smart Highway for the development of automated driving technology. Hyundai Motors was able to use the test environment on the Yeonju Smart Highway for its first truck platooning demonstration under real traffic conditions, using Xcient trucks. Efforts to digitalise the infrastructure are also being made in other countries. Among other things, Japan is working on preparing the infrastructure for a mixed traffic of truck platoons and conventional car traffic.

2.3 Conclusion

Truck platooning studies have shown that truck platooning facilitates increasing energy efficiency, may reduce costs and may improve road safety. Furthermore, studies provide information about socio-economic benefits such as congestion mitigation, traffic efficiency, better lane usage and driver safety. Although research and development as well as test results have been provided, the deployment remains open as well as further research and long-term impacts evaluations. Standards for multi-brand and multi-fleet platooning adopting C-ITS, clear platooning regulations and policies as well as safe and efficient operation strategies will be key for the adoption and deployment of truck platooning within the next years.

References

1. The European Automotive Manufacturers' Association (ACEA) (2017) What is truck platooning. https://www.acea.be/news/article/what-is-the-european-truck-platooning-challenge
2. European Union (2009) Promote Chauffeur II. https://trimis.ec.europa.eu/project/promote-chauffeur-ii
3. Fakultät für Maschinenwesen RWTHA (2009) Verbundprojekt KONVOI: Entwicklung und Untersuchung des Einsatzes von elektronisch gekoppelten Lkw-Konvois. Abschlussbericht
4. Chan E (2016) SARTRE Automated Platooning Vehicles. https://doi.org/10.1002/9781119307785.ch10
5. Eilers S (2015) Cooperative dynamic formation of platoons for safe and energy-optimized goods transportation. D3.2 Information Model for Platoon Services. https://cordis.europa.eu/docs/projects/cnect/0/610990/080/deliverables/001-D32Informationmodelforplatoonservices.pdf
6. ERTICO (2017) New Project on driving automation kick-off in Brussels. https://erticonetwork.com/new-project-driving-automation-kick-off-brussels/
7. Vissers J, Banspach J, Liga V, Tang T, Nordin H, Julien S, Martinez S, Villette C (2018) V1 platooning use-cases, scenario definition and platooning levels. D2.2 of H2020 project ensemble (platooningensemble.eu)
8. Schenker (2017) DB Schenker und MAN vertiefen Partnerschaft zum Autonomen Fahren. https://www.dbschenker.com/de-de/ueber-uns/presse-center/db-schenker-news/db-schenker-und-man-vertiefen-partnerschaft-zum-autonomen-fahren-13812
9. Schenker (2019) Platooning in der Logistikbranche: Forscher sehen nach Tests große Potenziale im realen Betrieb. https://www.dbschenker.com/de-de/ueber-uns/presse-center/db-schenker-news/platooning-in-der-logistikbranche--forscher-sehen-nach-tests-grosse-potenziale-im-realen-betrieb--594284
10. Menendez M (2018) Platoon - Einsatz von Lkw-Platooning-Strategien zur Verbesserung des Echtzeit-Verkehrsbetriebs auf der Autobahn. Netzwerk Stadt und Landschaft

11. Wang M, van Maarseveen S, Happee R, Tool O, van Arem B (2019) Benefits and risks of truck platooning on freeway operations near entrance ramp. Transport Res Record 2673:588–602
12. Bolen L (2019) Fahrermangel, Hyperloop oder Platooning? – Diese Themen werden 2019 den Transport beschäftigen. https://dispo.cc/a/fahrermangel-hyperloop-oder-platooning-diese-themen-werden-2019-den-transport-beschaeftigen
13. APA-OTS (2020) ASFINAG startet als erster Autobahnbetreiber Europas Vernetzung von Straße und Fahrzeug. https://www.ots.at/presseaussendung/OTS_20201020_OTS0067/asfinag-startet-als-erster-autobahnbetreiber-europas-vernetzung-von-strasse-und-fahrzeug-bild

Hatun Atasayar is an urban planner and researcher at the Austrian Road Safety Board (KFV) with focus on traffic safety and transportation. She works in several national R&I projects on the aspects of the digitalisation of the transport system with a particular focus on C-ITS and automated mobility.

Philipp Blass is a traffic safety researcher at the Austrian Road Safety Board (KFV) with a strong focus on automated driving and advanced driver assistance systems (ADAS). Within this field he contributes to several national and international research projects and aims to educate novice as well as experienced drivers in the topic of ADAS.

Susanne Kaiser is a psychologist and researcher at the Austrian Road Safety Board (KFV) since 2013. Her work is focused on human aspects in vehicle automation, driver state assessment, aggression in traffic as well as traffic safety culture. Susanne sis a member of HUMANIST VCE and the German Society of Traffic Psychology while also teaching Human Factors in Mobility at the University of Applied Sciences FH Technikum Wien.

Chapter 3
Towards Truck Platooning Deployment Requirements

Matthias Neubauer and Wolfgang Schildorfer

Abstract Truck platooning represents a promising means to enhance efficiency of freight transport. Developments of truck platooning date back to the early 1990s, starting with projects to illustrate the technical feasibility followed by projects investigating the potential of fuel savings up to feasible business models and multi-brand and multi-fleet platooning approaches. However, the deployment and adoption of truck platooning technologies need to detail, harmonise and finally meet diverse requirements. Figure 3.1 sketches requirement dimensions related to truck platooning, which span from safety and security requirements, stakeholder requirements to technical and functional requirements. In this chapter, selected results addressing requirements related to energy-efficient truck platooning, user and other road user requirements, road safety requirements and technical requirements related to C-ITS are presented.

Keywords Fuel saving · Truck driver · Regulation · C-ITS (cooperative intelligent transport systems)

3.1 Requirements Related to Energy Efficient Truck Platooning

Sustainable logistics demands for energy efficient transport of goods. In this context, energy-efficient transport needs to consider both (i) efficient and optimised transport plans/strategies as well as (ii) efficient transport execution (vehicles used, driving behaviour) itself.

Regarding efficient and optimised transport plans/strategies, [4] review published platoon planning approaches. In their contribution, they present three different planning situations based on the availability of truck travel information (start-/endpoint, departure-/arrival time):

M. Neubauer (✉) · W. Schildorfer
Department of Logistics Steyr, University of Applied Sciences Upper Austria, Steyr, Austria
e-mail: matthias.neubauer@fh-steyr.at

© The Author(s) 2022
A. Schirrer et al. (eds.), *Energy-Efficient and Semi-automated Truck Platooning*,
Lecture Notes in Intelligent Transportation and Infrastructure,
https://doi.org/10.1007/978-3-030-88682-0_3

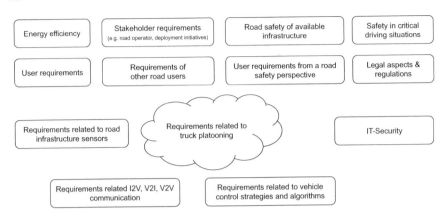

Fig. 3.1 Selected truck platooning requirement dimensions

- **Scheduled platoon planning**—truck travel information is available in advance and platoon plans may be developed and optimised off-line in a static way.
- **Real-time platooning**—truck drivers provide information when they start their trip [online or dynamic planning].
- **Opportunistic platooning**—truck that are nearby builds a platoon in a dynamic way without any plan in advance [ad hoc or on the fly platooning].

Bhoopalam et al. [4] state that truck platoon planning may serve different objectives, e.g. minimising system-wide fuel costs or maximising the number of platoons. Furthermore, efficient truck platoon planning needs to meet different requirements and constraints, e.g.:

- temporal coordination of truck journeys,
- regulations with respect to driving times,
- technical compatibility of platooning systems,
- constraints due to transport goods (e.g. dangerous goods),
- regulations related to platoon length,
- individual and organisational caveats,
- vehicle properties.

With respect to efficient transport execution, different influence factors need to be considered. Influence factors related to energy efficient vehicles are, e.g.:

- Drive technology

  – Increase efficiency of existing drive technologies versus alternative fuels.

- Vehicle weight.
- Aerodynamic drag.
- Rolling drag.

Furthermore, influence factors related to energy-efficient driving are:

- Driving behaviour, e.g.
  - driving foresighted, different driving manoeuvres for platoon building.
- Given (road) infrastructure, e.g.
  - Road profile, available real-time traffic information services.
- Current (road) traffic.
- Driving situation (wind, weather, ...).

Truck platooning as a means for increasing efficiency aims to reduce fuel consumption and emissions. Previous related work regarding truck platooning investigated potential savings gained by slipstream effects via reducing the distance between trucks and optimising the aerodynamic drag (cf. [2, 15, 17, 23])

Actual savings will depend on aspects such as distance between trucks in a platoon, overall truck platoon length/number of trucks in platoon, position of truck in platoon or platoon formation strategy/manoeuvre. Related work addressing energy efficiency of truck platoons applies simulation approaches (computational fluid dynamics, traffic simulation) and tests under real-life conditions on specific test tracks.

Simulation-based results reveal potential savings of truck platooning. For example, the EU Project Companion found based on a simulation model that even the first truck in a platoon may save 4.7–7.7% fuel at a speed of 70 km/h depending on the distance (10, 12, 20 m) between the first and the second truck (cf. [17]). However, tests on a test track showed within the Companion project that only minimal (max. 1%) or even no savings for the lead truck could be measured. [10] state based on the tests that the fuel saving potential is 4.99% per truck within a truck platoon consisting of three trucks driving at a speed of 70–80 km/h in a distance between 10 and 20 m. The companion tests were performed on a test track under ideal conditions. Effects such as current traffic or diverse platoon building manoeuvres were not investigated on the test track and will likely reduce fuel saving potentials.

Since the early 1990s, diverse truck platooning projects were realised (cf. 2). Estimating and evaluating energy efficiency of truck platoons typically represented a core element of such projects. Subsequently, reported fuel savings from different truck platooning projects are compared in Fig. 3.2 and the underlying data set is depicted in Table 3.1. The majority of the fuel efficiency investigations conducted tests on specific test tracks without surrounding traffic. Only the results reported by the German project EDDI (cf. [5]) were gained on German highways under practical operation. The listed investigations were performed for cabover trucks. The results can be compared with respect to the speed (in kilometre per hour—kph), the distance (in metre) between trucks in a platoon and the fuel savings for each truck in a platoon.

The comparison indicates different measured fuel savings across the projects. Differences may arise from different test situations (test settings, fuel measurement equipment, road profile, surrounding conditions). For example, the savings reported from tests under artificial conditions on test tracks are greater than those from tests under practical conditions:

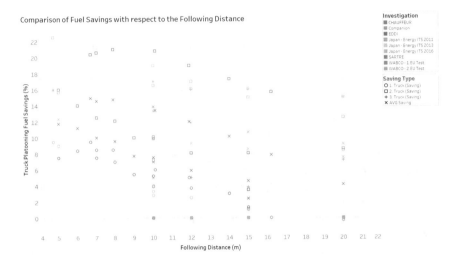

Fig. 3.2 Fuel savings comparison

- Saving results from Energy ITS [13, 21] at a distance of 15 m are in average 8.55%
  versus
- EDDI results (cf. [5]) gained under practical operation indicate fuel savings at a
  distance of 15 m in average of 2.4% for a platoon of 2 trucks.

However, valid (and agreed upon) criteria and data for assessing the energy efficiency
of truck platoons are vital for decisions makers to make an informed decision when
it comes to the deployment of truck platooning. Decisions in this case require multi-
criteria decision making taking into account different objectives and requirements
for truck platooning, for example defining reasonable distances between trucks that
are fuel efficient, still safe and comfortable for the truck driver.

Overall, the **requirements related to energy-efficient truck platooning** may be
summarised as follows. Energy-efficient truck platooning needs to consider:

- Platoon planning and formation strategies.
- Driving behaviours.
- Vehicle configurations, e.g.

  – max. speed, motor power, tyre pressure, vehicle weight incl. goods, braking
    profile, power train—diesel, liquefied natural gas, electric.

- Platoon configuration, e.g.

  – Truck sequence in platoon, distance between trucks.

- Road infrastructure.
- Control technology for efficient formation, execution and resolving of truck platoons.

Table 3.1 Fuel savings comparison—data set

Investigation	Year	Speed (kph)	Distance (m)	Savings (%)				Trucks in Platoon	Source
				1. Truck	2. Truck	3. Truck	AVG		
CHAUFFEUR	2000	80	6.7	9.50	20.40		14.95	2	[13]
CHAUFFEUR	2000	80	7	8.50	20.60		14.55	2	[13]
CHAUFFEUR	2000	80	7.9	8.50	21.00		14.75	2	[13]
CHAUFFEUR	2000	80	10.1	6.00	20.80		13.40	2	[13]
CHAUFFEUR	2000	80	11.9	5.00	19.00		12.00	2	[13]
CHAUFFEUR	2000	80	14	3.00	17.30		10.15	2	[13]
CHAUFFEUR	2000	80	16.2	0.00	15.70		7.85	2	[13]
Companion	2016	80	12	0.00	8.50		4.25	2	[10]
Companion	2016	80	15	0.00	7.80		3.90	2	[10]
Companion	2016	80	20	0.00	8.60		4.30	2	[10]
Companion	2016	80	12	0.00	9.90	5.00	4.97	3	[10]
Companion	2016	80	15	0.00	9.20	3.80	4.33	3	[10]
Companion	2016	80	20	0.00	9.00	7.20	5.40	3	[10]
EDDI	2018	80	15	1.3	3.5		2.40	2	[5]
Energy ITS 2011	2011	80	10	7.5	18	16	13.83	3	[22]
Energy ITS 2011	2011	80	10	10	17.5	14	13.83	3	[22]
Energy ITS 2013	2013	80	5	9.00	15.70		12.35	2	[13]
Energy ITS 2013	2013	80	10	3.30	16.50		9.90	2	[13]
Energy ITS 2013	2013	80	12	2.50	16.10		9.30	2	[13]
Energy ITS 2013	2013	80	15	1.00	16.10		8.55	2	[13]
Energy ITS 2013	2013	80	20	0.00	15.00		7.50	2	[13]
Energy ITS 2016	2016	80	4.7	9.5	22.5	16	16.00	3	[21]
Energy ITS 2016	2016	80	10	4	19	17	13.33	3	[21]
Energy ITS 2016	2016	80	12	2.5	17	16	11.83	3	[21]
Energy ITS 2016	2016	80	15	1	15	16	10.67	3	[21]
Energy ITS 2016	2016	80	20	0	12.5	15	9.17	3	[21]
WABCO-1.EU Test	01/2017	85	10	3.9	10.1		7.00	2	[13]
WABCO-1.EU Test	01/2017	85	20	−0.3	8.6		4.15	2	[13]
WABCO-1.EU Test	01/2017	85	30	0.7	7.6		4.15	2	[13]
WABCO-1.EU Test	01/2017	85	40	0.7	7.8		4.25	2	[13]
WABCO-1.EU Test	01/2017	85	50	0.8	8.3		4.55	2	[13]
WABCO-2.EU Test	08/2017	85	10	2.8	7.2		5.00	2	[13]
WABCO-2.EU Test	08/2017	85	20	−0.1	8.4		4.15	2	[13]
WABCO-2.EU Test	08/2017	85	30	0.4	7.1		3.75	2	[13]
WABCO-2.EU Test	08/2017	85	40	−0.4	7.1		3.35	2	[13]
WABCO-2.EU Test	08/2017	85	50	−0.2	7.4		3.60	2	[13]
SARTRE	2010–12	90	5	7.5	16		11.75	2	[13]
SARTRE	2010–12	90	6	8.4	14		11.20	2	[13]
SARTRE	2010–12	90	7	7.5	12.5		10.00	2	[13]
SARTRE	2010–12	90	8	7	12.1		9.55	2	[13]
SARTRE	2010–12	90	9	5.4	10		7.70	2	[13]
SARTRE	2010–12	90	10	5.2	9.9		7.55	2	[13]
SARTRE	2010–12	90	12	3.7	8.1		5.90	2	[13]
SARTRE	2010–12	90	15	1.1	8.1		4.60	2	[13]

3.2 User and Other Road User Requirements

In this section, requirements related, especially related to truck drivers and other road users, are detailed. The results are either derived via literature reviews and/or dedicated studies within the Connecting Austria research project.

3.2.1 Truck Driver-Related Requirements

Following, the results of a literature review related to truck driver requirements are presented. The literature search focused on:

- Technology acceptance + platooning.
- Human–machine interface (HMI) + platooning.
- Simulation + platooning.
- Truck platooning deployment.

During the literature search, Google Scholar as search engine and digital libraries (ACM, IEEE, Springer) were used to find relevant references. In addition, forward- and backward-oriented search strategies based on the literature found were applied. The results of the literature review are summarised in Table 3.2. The results are structured according to (i) Reference (Ref.), (ii) Study type, (iii) Study focus and (iv) Requirements/Issues identified.

In addition to the literature review, empirical studies on truck platooning acceptance in Austria were performed in Connecting Austria. In general, all participants of empirical studies conducted in Connecting Austria provided an informed consent to participate in the respective study and to publish the results. Furthermore, interview data were anonymised and securely stored internally in the University of Applied Sciences Upper Austria. In advance to the empirical studies, the authorisation of the involved parties related to the interviews was provided. In 2018, different empirical studies were performed together with Austrian fleet operators and their truck drivers with a special focus on the acceptance of level 1 platooning. In an initial endeavour, a questionnaire based on technology acceptance aspects presented by [24, 25] within the UTAUT and TAM model was conducted. In advance to the questionnaire, level 1 platooning was explained and demonstrated in a short presentation including texts, images and videos. Fifteen male truck drivers aged between 34 and 57 years participated in the questionnaire.

In a next step after the platooning acceptance questionnaire, semi-structured interviews with ten truck drivers aged between 33 and 54 years were conducted in order to gain deeper insights. The interviews were structured similarly to the questionnaire.

As reported in [16], the results of the questionnaire and the semi-structured interviews differ in some aspects. Compared to the questionnaire results, the interview results expect that platooning will be deployed within the next years successfully.

Table 3.2 Truck driver acceptance literature review

Ref.	Study type	Study focus	Requirements/issues identified
[14]	HMI study	Requirements related to truck platooning assistance systems and evaluation of design concepts and prototypes	−Acceptance −Comfort −Situation awareness −Loss of skill of truck drivers −Behavioural adaption −Risk compensation of drivers −Workload −Transitions from manual driving to automated driving −Response in case of breakdown −Accuracy and reliability of platooning systems is crucial
[14]	Driving simulator study	Explore opinions of drivers regarding gap sizes between platoon vehicles, problems related to driving near a platoon and perceived stress of drivers	Intra-platoon gap size: −The majority (75%) of the driver feel uncomfortable below 16 m gap size and unsafe below 7 m Driving near a platoon: −The majority (73%) of the driver feel that driving near a platoon of five cars and a leading truck is the same as normal driving
[19]	HMI design	Truck platooning HMI for performing platooning manoeuvres and increase situation awareness	Information related to own vehicle is more useful than information related to the overall platoon Information provision should be divided into three horizons: (i) close horizon −info related to own vehicle; (ii) medium horizon −info related to front and rear truck, distance, speed; (iii) far horizon −info on all trucks, common plan, future manoeuvres Truck drivers rate close and medium horizon information as most important Truck drivers expect that braking is always possible and sensors are working properly

Table 3.2 (continued)

Ref.	Study type	Study focus	Requirements/issues identified
[7]	HMI design	Presentation and evaluation of two novel human–machine interfaces for truck platooning	–Increase trust in system and reduce fear of drivers by appropriate system feedback and visualisation –The type of information visualisation influences trust in a system –The HMI design influences anticipating and avoiding of driving hazards –Assistance systems need to reach functional goals, e.g. hazard avoidance
[11]	Driving simulator study	Investigate the effect of partially and fully automated truck platoons on driver workload, trust, acceptance, performance, and sleepiness	–Workload for partial automation is higher than for full automation or manual driving –Trust in the system does not differ between partial and full automation –Drivers overestimate situation awareness during automated driving –Sleepiness increases in partially and fully automated platoons. –HMI needs to address workload, sleepiness, trust and legislation
[8]	Technology acceptance	Assessment of acceptance factors for future workplaces in highly automated trucks	–Regarding the overall acceptance, the current results indicate that truck drivers might not be in favour of highly automated trucks –Drivers' reservations regarding autonomous trucks may stem from uncertainty of system performance and uncertainty of the effect on employments and job profiles
[18]	Technology acceptance	Investigate truck drivers' and fleet managers' acceptance of highly automated driving to determine acceptance enhancing elements	–Safety and comfort are main reasons to use automated driving functions –Drivers fear to be made redundant and fear to loose driving pleasure –Truck drivers and fleet managers consider legal liability as well as the general safety and reliability of highly automated driving as crucial

Table 3.2 (continued)

Ref.	Study type	Study focus	Requirements/issues identified
[6]	Technology acceptance	Investigate key factors expected to influence truck platooning deployment	–Tailored use cases –Viable business models –Safety assurance –Consideration of human factors (interaction, change of roles and decision structures, training) –Public acceptance and trust –Impact on labour
[5]	Driver acceptance on –road experiment	Investigate truck platooning in real operation	–Driver perceives safety and driving comfort as main advantage of truck platooning –Drivers' experiences with platooning technology increase positive attitude towards safety and technology. Trust in platooning technology increased with experience –Vehicle cut-ins are more likely at a intra-platoon vehicle distance of 21 m. A distance of 15 m reduces vehicle cut-ins –The majority (90%) of the driver agreed on a distance of 15–21 m between trucks in a platoon –Truck driver stress measurements via EEG did not show a difference between normal driving and driving in a platoon
[20]	Driver acceptance on-road experiment	Investigate CACC in real driving situations	–The use of the CACC system in mixed public traffic was accepted by the truck drivers –An intermediate truck platoon gap setting was preferred by the truck driver –Truck driver did not report preferences related to 2nd or 3rd trucks in a platoon –The response of the CACC to cut-ins was accepted by truck drivers –The CACC system was disabled in case of high traffic density, large road grade and highway merging

Furthermore, the interview results show that drivers expect a benefit when using platooning, whereas the questionnaire results do rather indicate that drivers to not expect a benefit. Furthermore, the interview results related to the intention to use are far more positive than the questionnaire results.

Furthermore, an observation study was conducted in the Connecting Austria project to gain further insights in the everyday work of truck drivers. As reported in [16], an interesting initial finding is that experienced truck drivers tend to turn of driving assistance systems and state "I know how to drive, that's my job". Fleet operators also confirmed this finding. For this reason, it will be crucial to design proper deployment processes, actively inform about new features and explain their functionality as well as involve and train drivers when introducing platooning features.

Requirements related to truck drivers may be summarised as follows. Existing related work in the area of HMI provides insights in how interface designs for platooning should be designed and what are crucial acceptance factors, e.g. related to information provision or situation awareness. Furthermore, existing simulator studies and studies with research and development prototypes in real-world tests provide insights related to the application of platooning. These studies provide for example findings on acceptable distances between trucks, trust between truck drivers or trust in technology as crucial elements for deploying truck platooning. In general, the results of the empirical studies conducted in Austria in 2018 confirm the observations presented in the related work. However, there are some slight differences. For example, within the related work safety and comfort are identified as the main reasons for truck drivers to use automated driving functions (cf. [18]. The results of the empirical studies in the Connecting Austria project indicated that truck drivers do not think that platooning will increase safety. However, the truck drivers were asked before the actual use of such a system. Considering the findings within the real operation usage of platooning (cf. [5]) safety concerns could be eliminated during actual use. Beyond, the observation in the Connecting Austria project indicated that the general intention to use assistance systems may influence the adoption of truck platooning.

3.2.2 Other Road User-Related Requirements

The deployment of connected and automated vehicles will not appear from one day to another. Traffic situations in which vehicles at different automation levels and other road users need to interact are a major challenge for deploying automated driving solutions. Safe and smooth interaction among other road users and automated vehicles will play a crucial role when it comes to the acceptance of automated vehicles in real traffic situations.

Initial research regarding the communication between automated vehicles and other road users (e.g. pedestrians, bicyclists, SAE-level 0 vehicles) has already been presented. For example, [9] investigate "Automated Vehicle Interaction Principles" within four different scenarios:

1. A pedestrian encounters an automated passenger vehicle or a truck at a zebra crossing.
2. A pedestrian encounters an automated passenger vehicle at a parking lot.
3. A conventional passenger vehicle encounters an automated passenger vehicle or a truck at a symmetrically narrowed road.
4. A conventional passenger vehicle encounters a platoon of automated trucks on a highway.

Especially, scenario 1 and scenario 4 allow to convey requirements of other road users relevant in the context of truck platooning. Scenario 1 investigates the interaction between pedestrians and automated trucks at crossings, and scenario 4 is explicitly dedicated to truck platoons.

Related to the interaction at crossings [9] presents the following key requirements and challenges:

- Mutual understanding between pedestrians and automated vehicles needs to be supported. However,

    - Mutual understanding is challenged by the complexity of interactions and external as well as internal and situational factors, e.g. way of communication between system and other road user—explicit versus nonverbal communication to clarify intentions.
    - Today, pedestrians and drivers use diverse signals to negotiate mutual understanding, e.g. eye contact, hand waving, posture, lights flashing.

- Being aware of the vehicle mode (e.g. driving, stopping, accelerating) could allow pedestrians aligning their behaviour.
- Knowing the intentions of the vehicle would eliminate possible ambiguities due to the lack of communication with the "driver".
- Minimalistic and generic external vehicle interface such as AVIP could have positive impact on interactions between AVs and pedestrians.
- Ensure that pedestrians can interpret the signals displayed by an external vehicle interfaces.
- Ensure that external vehicle interfaces will function in a multi-agent scenarios.

To explore the needs of external signalling for truck platoons, [9] used various highway scenarios (on-ramp, off-ramp, etc.) as a starting point for a series of workshop discussions and interviews involving truck drivers with platooning experience as well as passenger vehicle drivers, OEMs and other experts in the field. Related to scenario 4 (conventional passenger vehicle encounters a platoon of automated trucks on a highway) [9] present the following key requirements and challenges:

- Inter-vehicle interactions are today affected mainly by traffic regulations. However, there traffic situations rely not only on regulations and depend on the perceptions and actions of traffic participants. In this respect, non-verbal communication aspects, such as eye contact or flashing the lights, need to be understood to be able to negotiate and align behaviour among traffic participants. Unfortunately, non-verbal cues may be ambiguous and situation dependent.

- Traffic safety and efficiency as well as trust towards AVs may be enhanced if AVs are able to communicate information on their state and information that "replaces" driver-centric cues such as hand gestures.
- External vehicle interfaces may be useful in solving ambiguities that drivers face today in unclear traffic situations.
- For trucks involved in a platoon, the rationale of external signalling is to inform other road users in the vicinity that platooning is ongoing and that the trucks wish to stay together without interruption.
- Most relevant situations for external signalling for platoons are highways at on-ramps, off-ramps, during overtaking and lane changes.
- Barrier to implement external signalling on the trailer is the fact that trailers are often switched between different haulers. It will be unclear who should cover the cost for the installation and maintenance of the signalling system on the trailers.
- The possibility of using existing positioning lights should be explored.

The research of [9] represents an initial approach to designing communication means and reveals the need for further research addressing the communication among automated vehicles and other road users. In addition, [3] performed a study of communication needs in interaction between trucks and surrounding traffic in platooning. They investigated "What information needs to be communicated from a platoon to the surrounding traffic to create safe and efficient interaction with the surrounding traffic"? As a result of workshops and interviews, they present a list of information items that may be relevant for exchange (cf. Table 3.3) between a platoon and the surrounding traffic.

In their study, [3] conclude that *"the in Sweden, and during traffic conditions similar to today's traffic, there is no urgent need to deploy external signalling for platoons with two trucks, neither if they are driving in ACC-mode with 1s headway or using V2V with 0.5s headway, since the traffic density in most cases will allow for other traffic to interact smoothly with platoons. In Europe, the situation is different because the traffic density is much higher. However, based on the current study, we can not conclude that external signalling is needed in Europe. The subjective data collected in the project support that cut-ins can be reduced to close to zero with a vehicle distance of 0.5 s headway.[1]"*

3.3 Road Safety Requirements

Road safety needs to be ensured when it comes to connected and automated driving solutions such as truck platooning. Thereby, road safety may investigate different objects of traffic systems, e.g.

- Safe road design.
- Safe vehicle design.

---

[1] © April 2021 Andersson, J., Englund, C., and Voronov, A., reproduced [3] with permission.

Table 3.3 List of communication needs identified by [3]. © April 2021 Andersson, J., Englund, C., and Voronov, A., reproduced with permission

Communication needs	Corresponding information
Cruising	
Need to say "We are a platoon" (generic)	The platoon companions show that they are together and drive as one unit
Need to know if you can trust a new platoon companion and to communicate that you are trustworthy	The joining platoon companion communicate its status and performance
On-ramp and Off-ramp	
Need to communicate when it is necessary for one vehicle or the whole platoon to change lanes	The platoon uniformly communicates its intention to change lanes
Need to show when and how to adapt speed to oncoming traffic	The platoon shows that it has recognised the oncoming traffic and early shows its intended manoeuvre
Need to communicate which vehicle in the platoon will give way to oncoming traffic	The vehicle shows that it has recognised the oncoming traffic and early shows its intended manoeuvre
Need to tell how to adapt vehicle speed, gaps and position to facilitate weaving with oncoming traffic	The platoon companions indicate its intention to open gap or not
Need for the oncoming vehicle to sync speed	The platoon indicates its relative speed to the oncoming traffic
Need for the oncoming vehicle to tell that "I expect you to let me in"	The oncoming vehicle communicates its presence, intention and request to join the highway
Overtaking	
Need to communicate time and distance to next off-ramp	The platoon communicates the distance and time to the next off-ramp to aid overtaking decisions
Need to know intention if other vehicle wants to cut-in (to let cars by)	The overtaking truck communicates its intention to let cars from behind pass
Roadwork	
Need to have a lead vehicle driver that is vigilant and can change lane in time when needed	The lead vehicle indicates its intention to change lane (if uniform lane changes, await acknowledgement from companions)

	Scenario					
	A	**B**	**C**	**D**	**E**	**F**
Scenario Description	"Still safe"	"Astra Study ¬Bridges"	"Astra Study"	"Play safe"	"Astra Study ¬Bridges¬Tunnels"	"Allow ramps"
Platooning Constraints						
Bridges (length > 100m)	+/-400m	n/a	b: 1200m, a: 1100m	+/-1000m	n/a	+/-400m
On-ramps (All)	+/-400m	b: 1200m, a: 1100m	b: 1200m, a: 1100m	+/-2000m	b: 1200m, a: 1100m	n/a
Off-ramps (All)	+/-400m	b: 1200m, a: 1100m	b: 1200m, a: 1100m	+/-2000m	b: 1200m, a: 1100m	n/a
Tunnel (All)	+/-400m	b: 2800m, a: 2600m	b: 2800m, a: 2600m	+/-1000m	n/a	+/-400m

Legend:
(n/a) - not applicable
(a:) - after:
(b:) -before:
(+/-) - after/before

Fig. 3.3 Truck platooning regulation scenarios

- Safe driving.

Previous studies already investigated potential opportunities and risks of truck platooning as well required technical and infrastructural conditions. For example, in Switzerland, the federal office for roads (ASTRA) published a feasibility study, which assess the potential and risks of truck platooning on Swiss highways (cf. [12]). This study concludes that the potential of truck platooning in Switzerland is limited due to the restriction elicited within the study and the high number of tunnels and road accesses in Switzerland. As depicted in Fig. 3.3, the study recommends to resolve truck platoons 1200 m ahead of motorway access points and form 1100 m after motorway access points (compare Scenario C). Furthermore, truck platooning in motorway tunnels is not recommended (resolve 2800m in advance, (re-)form 2600 after tunnel) and bridges are also not recommended. Regarding bridges, the study indicates the need for further research addressing acceptable maximum weight and road wear. Further related work on regulations will be described in Chap. **??** as well as the need for the harmonisation of regulations.

In the Connecting Austria project members of the Austrian Road Safety Board (KfV—Kuratorium für Verkehrssicherheit) assessed requirements and regulations with respect to safe truck platooning and defined the following constraints within the project based on literature research, workshops and interviews with road safety experts:

- Motorway should comprise three lanes.
- Width of lane should be greater than 3 m.
- Platooning should be prohibited in tunnels (resolve platoon min 1000 m ahead).
- Platooning should be prohibited at road works (resolve platoon min 1600 m ahead).
- Platooning should be prohibited at declared dangerous spots (e.g. foggy areas, accident accumulation points).
- Platooning should be primarily performed on road sections with little curves.
- Platooning should be prohibited at motorway junctions (resolve platoon min 2000 m ahead).
- The length of acceleration lanes and deceleration lanes needs to be assessed.

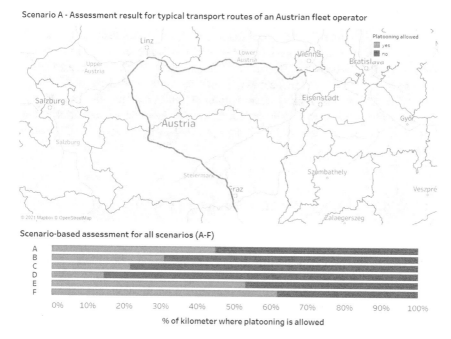

Scenario A - Assessment result for typical transport routes of an Austrian fleet operator

Scenario-based assessment for all scenarios (A-F)

Fig. 3.4 Scenario-based assessment of truck platooning routes for a fleet operator, base map and map data from OpenStreetMap, ©OpenStreetMap contributors under the CC-BY-SA license, https://www.openstreetmap.org/copyright

Scenario D in Fig. 3.3 adheres to the defined constraints of the KfV. In addition to this general regulations, the KfV recommends to assess road safety for dedicated platooning tracks. An established means to assess road safety is the "road safety inspection" method. The method has been adopted for truck platooning within the Connecting Austria project. The adoption as well as method application for a dedicated route will be described in Chap. ??.

In order to assess the feasibility of truck platooning in Austria, additional scenarios next to the rather conservative constraints defined in scenario B,C,D were defined by the project partners.

Scenarios A, E and F are aimed at maximising the amount of "allowed" truck platooning kilometres on Austrian motorways applying different strategies:

- Scen A—minimises area before and after potential dangerous spots to ±assumes that platoon may be safely operated on bridges and in tunnels and allows platooning on bridges and in tunnels (other parameters are similar to Astra study).
- Scen F—assumes low traffic density at motorway access points (e.g. during night) and allows to drive in a platoon at this road sections.

The scenario restrictions are summarised in Fig. 3.3. Furthermore, a selected result of the scenario-based assessment of routes is depicted in Fig. 3.4. The assessment

depicts typical routes of an Austrian fleet operator and the share of highway kilometres that are eligible for truck platooning within a certain scenario. The map in
Fig. 3.4 depicts the results, especially for scenario A. However, the chart below the
map depicts the differences between the defined scenarios for the given routes. A
detailed route analysis with respect to fuel efficiency will be described in Sect. ??.
In order to increase the number of in-platoon-driven-kilometres, dynamic platoon
management strategies are recommended. A situation-based implementation aiming
at dynamic platoon management on highways may be supported by C-ITS services
(see Sect. 3.4).

3.4   Technical Requirements Related to C-ITS

Intelligent road infrastructures fuelled by C-ITS services may have a positive effect
on the deployment of truck platooning with respect to safety and energy—as well as
traffic-efficiency (cf. [1]). In the area of C-ITS communication, standards have been
developed by ETSI. ETSI, a European Standards Organisation, already standard-
ises communication means for Vehicle-to-Vehicle (V2V), Vehicle-to-Infrastructure
(V2I) and Infrastructure-to-Vehicle (I2V) communication. Information exchange is
standardised in ITS-G5 messages, which may have different types:

- Cooperative Awareness Message (CAM).
- Decentralised Environmental Notification Message (DENM).
- In-Vehicle Information (IVI).
- Signal Phase and Timing (SPAT).
- MAP Topology Layer (MAPEM).

In the Connecting Austria project, relevant messages have been identified based
on the defined truck platooning use cases. For such messages, the data profiles need to
be defined. General information relevant for supporting truck platooning comprises:

- Information for building a platoon (I2V).
- Information for resolving a platoon (I2V).
- Information related to dangerous areas (e.g. road works)(I2V).
- Information related to the signalling of traffic lights (e.g. remaining green time)
  (I2V).
- Information related to road profile (e.g. traffic lights on a track) (I2V).
- Information related to the status of a platoon (V2I).
- Information exchange within a platoon (V2V).

For forming and dissolving platoons, the following specific messages are relevant:

- **Minimal gap distance** message—indicates to allowed distance between trucks in
  a platoon for a certain area. May depend on weather, traffic situation, regulations,...
- **Platooning allowed** message—indicates if platooning is allowed in a certain area.
  May depend on weather, traffic situation, regulations,...

Such C-ITS messages will allow to support dynamic truck platoon management for road operators. Depending on a given situation, the minimal gap distance may be varied or platooning may even be prohibited. In comparison with static and restrictive platoon regulations, a dynamic truck platoon management could increase the number of platooning kilometres on specific motorways and thereby contribute to energy efficiency and sustainability. However, information security of ITS-G5 communication needs to be ensured in terms of:

- Validation of message integrity.
- Validation of message authenticity.
- Availability of communication channels.
- Availability of communicating actors (road-side units and on-board units).

3.5 Conclusion

In this chapter, selected deployment requirements for truck platooning are reported. Starting with an overview on related work with respect to energy efficiency, the results of fuel savings investigated in previous projects are compared and different strategies to form truck platoons are explained. Subsequently, truck driver-related requirements are summarised based on a literature review and empirical studies conducted in Connecting Austria. Furthermore, user requirements related to other road users are compiled based on related work. In addition to user requirements, road safety requirements are derived and relevant scenarios of the Connecting Austria project are depicted. Finally, technical requirements relevant for C-ITS-based truck platooning are described.

References

1. Agriesti S, Gandini P, Marchionni G, Paglino V, Ponti M, Studer L (2018) Evaluation approach for a combined implementation of day 1 c-its and truck platooning. In: 2018 IEEE 87th vehicular technology conference (VTC Spring). IEEE, pages 1–6
2. Al Alam A, Gattami A, Henrik Johansson K (2010) An experimental study on the fuel reduction potential of heavy duty vehicle platooning. In: 13th International IEEE conference on intelligent transportation systems, pages 306–311
3. Andersson J, Englund C, Voronov A (2017) Study of communication needs in interaction between trucks and surrounding traffic in platooning
4. Bhoopalam AK, Agatz N, Zuidwijk R (2018) Planning of truck platoons: a literature review and directions for future research. Transp Res Part B: Methodol 107:212–228
5. Brandt A, Jentzsch G, Pradka A, Eddi electronic drawbar - digital innovation project report - presentation of the results
6. Engström J, Bishop R, Shladover SE, Murphy MC, O'Rourke L, Voege T, Denaro B, Demato R, Demato D (2018) Deployment of automated trucking: challenges and opportunities. Automated vehicles symposium 2018:149–162

7. Friedrichs T, Ostendorp M-C, Lüdtke A (2016) Supporting drivers in truck platooning: Development and evaluation of two novel human-machine interfaces. In: Proceedings of the 8th international conference on automotive user interfaces and interactive vehicular applications, automotive'UI 16, New York. ACM, pages 277–284

8. Fröhlich P, Sackl A, Trösterer S, Meschtscherjakov A, Diamond L, Tscheligi M (2018) Acceptance factors for future workplaces in highly automated trucks. In *Proceedings of the 10th International Conference on Automotive User Interfaces and Interactive Vehicular Applications*, AutomotiveUI '18, New York. ACM, pages 129–136

9. Habibovic A, Andersson J, Malmsten Lundgren V, Klingegård M, Englund C, Larsson S (2019) External vehicle interfaces for communication with other road users? In: Road vehicle automation 5. Springer, pages 91–102

10. Hanelt A, Hildebrandt B, Leonhardt D, Companion d8.4. documentation on the socio economic impacts of the project results

11. Hjälmdahl M, Krupenia S, Thorslund B (2017) Driver behaviour and driver experience of partial and fully automated truck platooning–a simulator study. Eur Transp Res Rev 9(1):8

12. Jermann J, Oehry B, Bosch R, Schmid T, Gasser Y, van Driel C, Kryeziu G, Chancen und risiken des einsatzes von abstandshaltesystemen sowie des platoonings von strassenfahrzeugen - machbarkeitsanalyse: Bericht-nr. 2060.966-001

13. Kallenbach S (2019) Truck platooning—a pragmatical approach. In: Fahrerassistenzsysteme 2018. Springer, pages 132–157

14. Larburu M, Sanchez J, Rodriguez DJ (2010) Safe road trains for environment: human factors' aspects in dual mode transport systems. In: ITS world congress, Busan, Korea

15. Liang K-Y, Mårtensson J, Johansson KH (2013) When is it fuel efficient for a heavy duty vehicle to catch up with a platoon? IFAC Proc 46(21):738–743

16. Neubauer M, Schauer O, Schildorfer W (2019) A scenario-based investigation of truck platooning acceptance. In: International conference on applied human factors and ergonomics. Springer, pages 453–461

17. Pillado M, Garcia-Sol L, Batlle J, Sanchez S, Freixas A, Shadeghian S, Friedrichs T, Companion d7.1: Limited results of the on-board coordinated platooning system performance evaluation via physical testing

18. Richardson N, Doubek F, Kuhn K, Stumpf A (2017) Assessing truck drivers' and fleet managers' opinions towards highly automated driving. In: Advances in human aspects of transportation. Springer, pages 473–484

19. Sadeghian Borojeni S, Friedrichs T, Heuten W, Lüdtke A, Boll S (2016) Design of a human-machine interface for truck platooning. In: Proceedings of the 2016 CHI conference extended abstracts on human factors in computing systems, CHI EA '16. ACM, New York, pages 2285–2291

20. Shladover S, Lu X-Y, Yang S, Ramezani H, Spring J, Nowakowski C, Nelson D, Cooperative adaptive cruise control (cacc) for partially automated truck platooning

21. Tsugawa S, Jeschke S, Shladover SE (2016) A review of truck platooning projects for energy savings. IEEE Trans Intell Veh 1(1):68–77

22. Tsugawa S, Kato S, Aoki K (2011) An automated truck platoon for energy saving. In: 2011 IEEE/RSJ international conference on intelligent robots and systems. IEEE, pages 4109–4114

23. van de Hoef S, Johansson KH, Dimarogonas DV (2015) Fuel-optimal centralized coordination of truck platooning based on shortest paths. In: 2015 American control conference (ACC), pages 3740–3745

24. Venkatesh V, Bala H (2008) Technology acceptance model 3 and a research agenda on interventions. Decis Sci 39(2):273–315

25. Venkatesh V, Morris MG, Davis GB, Davis FD (2003) User acceptance of information technology: toward a unified view. MIS Q 27(3):425–478

Matthias Neubauer has received her Masters Degree in Mechanical Engineering from theVien-naUniversity of Technology. Her research interests include traffic modelling, simulation and control, intelligent transportation systems and automation. Elvira has worked on national research projects and her academic contributions are published in relevant journals and conferences.

Wolfgang Schildorfer is a person who likes the road he still walks to find new chances. Since October 2018, he has been Professor for Transport Logistics and Mobility at the University of Applied Sciences Upper Austria. His research focus is on innovation, business models and evaluation in transport logistics, smart hyperconnected logistics systems, (urban) mobility, sustainable transport systems and new technology markets (C-ITS, CCAM, automated driving, truck platooning).

Chapter 4
Research Design and Evaluation Strategies for Automated Driving

Andreas Kuhn, José Carmona, and Elvira Thonhofer

Abstract Automated driving, in general, and platooning, in particular, represent a highly active field of research. The idea to automate traffic is closely related to high expectations in both individual and public transport. However, the complexity of automated driving requires methods beyond the traditional development approaches. This chapter describes a state-of-the-art methodology to organise and systematically address a comprehensive set of research questions in the context of truck platooning. Following best practices, an evaluation design is presented, which ensures the alignment of research efforts with the actual research agenda, that is, to answer the right questions. Specifically, the benefits of automated driving and their conflicting relationships are explored and the entities that affect automated driving performance and their interactions are presented. Finally, a solution concept that adequately addresses the complexity and the stochastic nature of the problem is presented. The solution concept consists of several key methods such as scenario-based design and stochastic simulation, data mining and complexity and robustness management.

4.1 Benefits of Automated Driving

The expected benefits of automated driving, and of platooning in particular, can be grouped into the following five categories:

- Comfort.
- Safety.
- Traffic efficiency.
- Traffic effectiveness.
- Vehicle efficiency.

A. Kuhn (✉) · J. Carmona · E. Thonhofer
Andata Artificial Intelligence Labs, Hallein, Austria
e-mail: andreas.kuhn@andata.at

© The Author(s) 2022
A. Schirrer et al. (eds.), *Energy-Efficient and Semi-automated Truck Platooning*,
Lecture Notes in Intelligent Transportation and Infrastructure,
https://doi.org/10.1007/978-3-030-88682-0_4

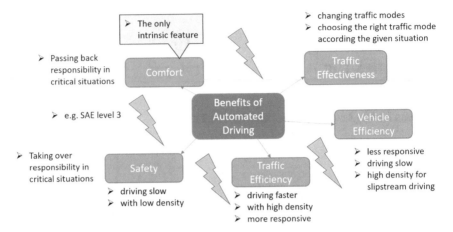

Fig. 4.1 Conflicting benefits of automated driving

Comfort benefits often translate directly to *time* that the driver of a vehicle can spend on tasks *other than* driving. Instead, travelling time can be used for tasks with significantly more added value or even be spent resting, similar to longer train rides.

Safety benefits aim at elimination or at least mitigation of collisions. Currently, up to 90% of all collisions are caused by human error, which may be eliminated by automation and automated driving functions.

Traffic efficiency benefits fall into two subgroups. First, it is expected that traffic volume across an intersection or through a section of highway can be increased while keeping travel time low and avoiding congestion. Second, it is expected that automated vehicles can be operated at superior efficiency compared to human-driven vehicles, due to predictive driving strategies and cooperative behaviour.

Vehicle efficiency considers energy consumption and energy efficiency. For example, predictive driving strategies may improve vehicle efficiency by saving fuel.

Traffic effectiveness benefits can be achieved by selecting the best mode of traffic in any given situation. In case the road network of the city centre is already congested, the capacity of the system is better utilised if additional journeys are undertaken by metro instead of individual cars. Shared mobility is also a means of utilising this benefit.

These groups are connected, and many benefits are conflicting; see Fig. 4.1.

Each benefit category can be investigated individually; however, due to their conflicting nature, the performance in every other category will be poor. The only intrinsic benefit of automated driving is comfort. Therefore, the benefits of all categories must be managed well in order to achieve some sort of satisfactory overall performance. Mathematically speaking, the Pareto front of the benefit criteria (also known as key performance indicators—KPIs) has to be evaluated and properly weighted solutions at the edge of the Pareto front have to be found. In the subsequent sections, different requirement conflicts of the above-mentioned benefits of automated driving are discussed.

4.1.1   Requirements Conflict Efficiency Versus Safety

Looking at the defining equation for traffic volume $Q = D \cdot V$ and recalling the fundamental diagram, traffic density D and/or traffic velocity V have to be increased in order to increase the flow rate Q per lane. Higher velocities and higher traffic density correlate positively with larger traffic volume. However, increasing safety in traffic usually translates to reduced velocity and/or lower traffic density. Reducing the velocity reduces collision severity, and reducing traffic density reduces the risk of collision.

4.1.2   Requirements Conflict Safety Versus Comfort

Most current advanced driver assistance systems (ADAS) are comfort systems by design. They are designed to operate under regular conditions. As soon as a critical traffic situation is reached, these systems delegate responsibility back to the human driver. In contrast, safety systems are designed to operate in critical situations, where these systems assume control. The human driver does not recognise that a safety system is operating under regular driving conditions. Hence, safety systems and common comfort enhancing assistant functions operate reciprocally. The importance of handover scenarios in automation systems, in particular at SAE level 3, must be discussed and addressed by the scientific community .

4.1.3   Requirements Conflict Comfort Versus Effectiveness

Significantly increasing driving comfort and allowing productive tasks being performed while travelling by car may cause even more individual vehicle traffic on road networks. This will influence the choice of traffic mode, in addition to many other aspects that a person considers while making this choice.

4.1.4   Requirements Conflict Comfort Versus Efficiency

Public transport is undoubtedly more efficient in terms of passenger per section and time. However, comfort levels of public transport vary with mode type, time of day (crowdedness and cleanliness), particular route requirements, etc., and are generally lower than when travelling in one's own car.

4.1.5 Requirements Conflict Traffic Versus Vehicle Efficiency

Traffic and vehicle efficiency are in conflict too. While traffic efficiency sometimes requires high responsiveness and higher accelerations to close distance gaps and still reaches remaining slots of green times, vehicle efficiency normally needs smooth changes in velocities and slow, energy preserving motion patterns for certain street sections and traffic events, like red lights and accelerations after changes to green light.

Summing up, the mentioned main categories of benefits from automated driving are conflicting and oppositional. Comfort is the only intrinsic feature, which comes along on its own purely by the introduction of vehicle automation. A sole automation of vehicles only will not automatically make traffic safer, more efficient and more effective. These properties must be worked out cumbersomely if all the expected benefits of automated driving should be leveraged and be put into real effect. Hence, it is not a matter of optimisation, potential benefits have to be balanced and proper trade-offs have to be found between the counteracting effects. This can only be done by gathering knowledge about the quantitative relations of the described effects. Mathematically speaking, the Pareto front of the benefit criteria (i.e. the KPIs) must be evaluated and properly weighted solutions at the edge of the Pareto front have to be found.

4.2 Entities with Effects on Automated Driving Performance

When searching a quantitative description of the various performance parameters and dependencies of automated driving effects, the relevant parameters and entities have to be identified first. Figure 4.2 shows the main entities which have influence on the performance of automated driving functions and have to be considered, in case the potential benefits of automated driving should be put into effect. These entities must be described and elaborated numerically. Unfortunately, all these entities have an effect on all the others and, therefore, must be considered simultaneously.

ADAS and Automated Driving Functions: Of course, the realisation and certain characteristics of the advanced driver assistance systems and automated driving functions themselves have the primary influence on the benefits of automated driving. The way they brake, steer, manoeuvre and execute their missions has fundamental influence on the re/action of other traffic participants and the traffic situation. The more precise the control algorithms estimate, predict and consider the behaviours of other traffic participants, the better their control performs. Telling other traffic participants and traffic control systems about the own intention respectively being deterministic allows others to better cooperate for optimal overall control.

Traffic Control and Management Systems: When optimising traffic is really a goal for automated driving, not only single vehicles but also traffic control and management should be automated to take the new functional possibilities and their potential into account. Traffic control can be improved by vehicle automation because the behaviours of the automated cars become more deterministic and predictable. Communication of the vehicles and traffic participants' intentions also helps intensively for the interpretation of the traffic situation to adapt optimal control strategies. On the other side, automated vehicles are expected to fulfil advices from traffic control more disciplined and more precisely than humans would do. Of course, such "advices" have to be described properly by laws, regulations and common standards. Being clever in defining such a specification can help improving the overall system performances dramatically.

Behaviour of Drivers and Traffic Participants: Traffic participants' behaviour determines heavily the resulting traffic situation. The more predictable and cooperative the traffic participants behave, the more precise and better vehicle and traffic control systems can become. Automation also allows prescribing beneficial behaviours more strictly, which normally will also be followed then. Automated vehicles also need to be predictable in their behaviour for non-automated traffic participants. Anyhow, this issue deals with all the effects of mixed traffic of automated and non-automated traffic participants and their (hopefully) cooperative interactive behavioural patterns. Due to the fact that there will always be mixed, individual traffic, the interactive behaviour of automated and non-automated traffic participants has to be considered anyway, as long as they are not strictly separated (what will be very difficult, especially in urban environments).

Traffic Situation: The traffic situation is mainly influenced by the behaviour of the traffic participants and the traffic and vehicle control strategies. The identification of the traffic situation is also the main control parameter for optimal, adaptive traffic control.

Streets and Infrastructure: Infrastructure—when properly maintained—can simplify the control task dramatically (e.g. by clear and consistent markings).

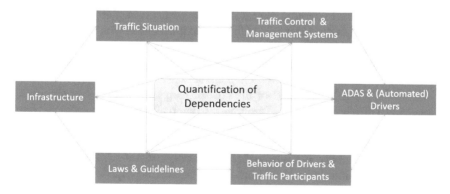

Fig. 4.2 Main entities with influence on the performance and benefits of automated driving

On the other hand, vehicle control must also be able to coup with imperfect and disturbed environments.

Laws, Regulations and Guidelines: The main role of regulations and laws are defined by the "specification" of the necessary behavioural patterns and technical standards, especially with a priority on safety issues and cooperative control policies.

The mentioned relations between the main entities are not complete. Even more have been identified within the project WienZWA [3]. Due to the complexity and the interdisciplinary aspects of automated driving, the whole issue of vehicle automation cannot be resolved by single parties (e.g. a vehicle manufacturer). Costs and expenses would be much higher compared to collaborative approaches. Each of the described entities also corresponds to different stakeholders and scientific communities with different backgrounds. Bringing them together is not only a technical but also an organisational challenge. Technically, such a collaboration is required. Vehicle and traffic automation means that driving and traffic control is mainly substituted by algorithms. Therefore, if these algorithms should work optimally, they require precise quantitative information about the other entities. Furthermore, other parties need to get a feeling how their contribution influences the others. In the opinion of the authors, the best organisational approach for such a common playground would be an open test field, where all the described entities could be tested in common and brought into the overall context.

4.3 Additional Sources of Complexity

In the previous section, it has been illustrated that automated vehicles are part of an overall, much more complex (control) system-of-systems, when looking at the aspects of traffic and cooperative driving instead of only looking at single vehicles. This system-of-systems and all its relevant effects on the automated cars have to be taken into account, when validating automated vehicles and proofing their positive effects for improved traffic. Practically, such validation can only be done under real conditions in an open test field with realistic interactions with other traffic participants. But according to [7], there are more reasons, which motivate test fields and accompanying advanced validation methods, even when purely looking at the problem from the single vehicle system point of view.

First of all, humans are part of the control loop now, not only externally by interacting with other traffic participants but also internally. Hence, control actions are not only rational and deterministic than anymore. Many decisions in traffic are based on the subjective judgement from traffic participants with very different experiences, backgrounds and skills. For the overlaying control, it is not a matter of objective system identification, but rather about estimating and anticipating the difference between objective and subjective estimations of risks and consequences. But not only humans and their partly irrational re/actions in certain situations are a source

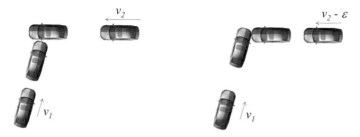

Fig. 4.3 Instability due to initial conditions for collisions

of stochastic variability. In addition, also sensors and actors are still far away from being perfect and precise. Besides the uncertainty of the sensor signals, also the classification performance of the algorithms may lack performance. The problem is stochastic; therefore, a probabilistic point of view with according methods is an absolute must. Furthermore, many decisions for the choice of driving strategies are context sensitive. They rely of information, which cannot be sensed directly. And last but not least, the problem is unstable. Even very small changes may lead to completely different results. For example, when looking at Fig. 4.3, dependent on very small differences in the initial velocity of the displayed configuration, the results for one of the cars may be a front crash or a side crash in the other case. From the perspective of automotive safety, these are completely different incidents with very different consequences and necessary countermeasures. On the other side, also very small measures at the right time may help to avoid such critical situations at all. Unfortunately, also the opposite is true. Very small errors or fail behaviours may result in severe crashes or heavy congestions.

4.4   Development Procedures

In Sects. 4.2 and 4.3, it has been shown that the validation and rating of automated driving is a very complex task with a lot of interfering disciplines and stakeholders. Therefore, traditional knowledge-driven development methods and procedures are not sufficient anymore and have to be extended by modern approaches, which leverage the possibilities of new technologies from connectivity, big data analytics, artificial intelligence, etc., as fundamental part of the development and validation procedure.

Under knowledge-driven development, the linear, sequential development procedure is meant, which is roughly shown at the top of Fig. 4.4. Based on the knowledge and experience of certain domain experts, a solution concept is drawn and implemented. After that, the implementation is assessed in tests and under real conditions. Dependent on the time duration of that procedure and the complexity of the problem, the assessment often results in the finding that the assumptions of the experts and

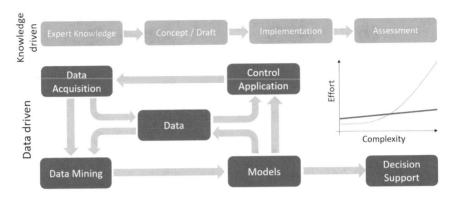

Fig. 4.4 Knowledge-driven development versus data-driven development procedures

requirements from the beginning of the development are not valid anymore. The development has to be repeated or heavily adapted. In the case of automated driving, this issue is even intensified. According the different entities from Sect. 4.3, several very different disciplines and fields of expertise are contributing to the topic. There will not be any experts with sufficient knowledge in all of the necessary disciplines and subtopics. Therefore, lots of experts from different domains and organisations have to be coordinated and brought together. Also, each of the topics is running in completely different timescales. While mobile development (which is used, e.g., for traffic information systems) is running in innovation cycles of months, automotive development is running in cycles of years, and infrastructure issues are a matter of decades. Therefore, the synchronisation of the contributing different branches for automated driving will be quite challenging.

A major consequence would be the implementation of a common playground in the form of open test fields, which cover all the described entities. The technical coordination and integration will be done in the form of a data-driven development approach. The principle schema is shown in Fig. 4.4. Basic element is a data pool, where all acquired data from the test field, from naturalistic driving, from dedicated tests and from various simulations is collected and composed. Steadily acquired new data is used to update all the mathematical models and algorithms, which are part of the control application. Data mining and Machine Learning methods are used for continuous adaptation of the models. The results of the updated control applications are again acquired and fed back to the basic data set. In that way, an environment is defined, which allows steady iterations and continuous updates from instantaneously learned results. Such a procedure can also be seen as an agile system and functional development procedure, which is inspired by agile software processes. These allow extremely quick development cycles especially in case of complex systems, where precise requirements and specifications are not available from the beginning. The technological basis is the described solution components from the next chapter as well as connected technologies and Internet of Things solutions, which allow a dramatic acceleration of such learning cycles. Hereby, all the systems change because behavioural changes are captured immediately after application of the new solutions, allowing a quick adaption in case of unexpected effects and changes. In [1, 2, 6], it has been shown that such data-driven product development approaches are much quicker and much more effective and efficient in the case of complex systems than traditional purely knowledge-based approaches.

4.5   Solution Concept

When validating the effects and benefits of automated driving, first of all each of the described entities from Sect. 4.2 and the corresponding parameters and measurement should be available in a test field for automated driving. Laws and regulations are assumed to be quasi-static and available anyhow. Of course, these also need to be available explicitly, when testing policies and regulations.

But there are some reasons why the sole observation, tracking and logging of automated cars are not sufficient to validate and rate the effects of automated driving efficiently. Due to

1. the stochastic nature and
2. the complexity

of the problem, it would take a lot of test cycles and time to find out and proof the real reasons for the various effects. Therefore, the combination of traffic observation in the test field with some accompanying advanced methods is recommended to increase learning rates dramatically.

Following the data-driven approach from Sect. 4.4, a comprehensive scenario catalogue with numerical simulations is built up, maintained and steadily extended in parallel to the test field. The resulting scenario database covers all relevant cases from the test field and even goes far beyond. The scenario database contains simulation data from all relevant street sections of the test field in detail. In addition, numerous other cases from outside the test field are part of that scenario database. Results from the test field are immediately projected to the scenario database. Accordingly, applied methods to the scenario database then help in the interpretation and generalisation of the test results. The scenario database also includes the evaluation of criteria, which cannot be measured directly, like collision and/or congestion risks. An excerpt of these methods is described in the following. For all the mentioned methods at least, a proof of feasibility is available from the authors. For most of the methods already mature solutions are available. They just have to be combined accordingly and put together for a given test field.

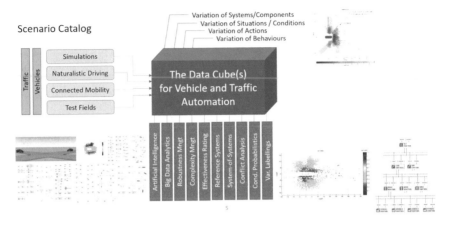

Fig. 4.5 Scenario-based approach for the development and validation of vehicle and traffic automation

4.5.1 Scenario-Based Approach and Stochastic Simulation

The basic method accompanying the test field runs is a scenario-based approach for the representation of all possible variations due to traffic and vehicle automation, like it is illustrated in Fig. 4.5. Hereby, a comprehensive collection of simulation scenarios is built up including variations of

- The infrastructure (e.g. geometry and topology of streets and intersections, road conditions, lane configurations, …).
- Traffic situations.
- Traffic participants and their behaviours.
- Traffic control strategies.
- Vehicle control strategies and control actions.

The variations are done with the help of multi-layered stochastic simulations (also known as Monte Carlo simulations) including the consideration of conditional probabilities (Bayesian approaches). The distributions for the Monte Carlo simulations are taken from the observations in the test field and from naturalistic driving studies, assuring representativeness of the data. The scenario catalogue contains the test field itself in detail. Furthermore, any other kind of street sections and intersection layouts are collected and composed in the scenario catalogue. These are built up automatically with the help of geoinformation systems like the Austrian graph integration platform (GIP), OpenStreetMap or similar other sources. The traffic situations are varied with respect to the real distributions of traffic situations taken from observations. Doing so, not only traffic volumes are varied, but also the distribution of the traffic participants and their behaviour is varied. The distributions are taken from external observations (e.g. by video tracking) or from their internal tracking with naturalistic driving/riding/walking data. Further possible control actions for traffic control as well as for vehicle control are varied. For these, special care has to be taken, according to [7], to assure the validity of the data and its interpretation. Once a test run is done in the real test field, the observations are mapped to the scenario catalogue. Thereby a single observation is put into the context of the occurring stochastic variations and can be classified as either a regular or an abnormal incident. With the help of the variations for the control actions in the scenario catalogue, also the consequences of the application of alternative driving control strategies and traffic control policies can be evaluated immediately. When observing that an event/situation is not yet represented in the scenario catalogue, the catalogue will be extended by that new situation. The scenario catalogue does not only contain driving situations. It contains different levels of details, ranging from vehicle dynamic simulation via traffic micro-simulation to macro-simulation. These different hierarchy levels of control are combined according to [4, 5]. Special multi-layered stochastic simulations and hierarchical approaches (some are betoken in [2, 5, 7]) allow the evaluation of realistic collision probabilities, criticality rates, collision risks, reserves with respect to safety, traffic volumes, etc. Many of such not directly measurable values are calculated in the scenarios catalogue and used later for the effectiveness rating (see Sect. 4.5.5) or the development of virtual sensors (see Sect. 4.5.6).

4.5.2 Big Data Analytics and Machine Learning

When having real big data in the scenario database and the test runs, it is obvious to approach this data with suitable, modern data mining, respectively, machine learning methods. These methods help to identify dependencies and relations between the relevant entities, parameters and influence factors. It is also the method of choice for the identification and predictions of the traffic situations. Once having captured all the relevant effects in the scenario catalogue, the resulting machine learning models can be used as control models as well, for instance, to estimate/predict the behaviours of the other traffic participants, for traffic control or for the selection of the proper driving and traffic control strategies according to the given traffic situations. Because the machine learning models can be adapted and updated quickly after new findings from the test runs, these are also the appropriate method to incorporate the learning from the test field results.

4.5.3 Incident and Anomalies Detection

Incident and anomalies detection is a generic method for quality assurance of any kind of data. It can be used for several use cases within the test field:

- The identification of abnormal situations helps to fill up the scenario catalogue with all kind of relevant traffic situations.
- New control concepts can be cancelled immediately if they result in surprising or unfavourable effects.

- The quality of the simulation and test data can be assured, by sorting out false and buggy data.

4.5.4 Naturalistic Driving and Behavioural Models

Naturalistic Driving is the common keyword for the evaluation of driver behaviour through observation of drivers under naturalistic conditions in real traffic. That way, the behaviour of drivers with respect to the environmental conditions in different traffic situations can be assessed. The resulting models can be used for the automated driving functions to anticipate the behaviour of the other traffic participants. Further on, these models can be used in the scenario database to evaluate the consequences of a different behaviour of the traffic participants. When using naturalistic driving in a test field with the according sensor measurements, the method can also be used as usability test for infrastructure and traffic control systems. In that sense also, the acceptance of control strategies can be evaluated.

4.5.5 Effectiveness Rating

Automated driving and traffic automation is a new field, where still the best control actions and strategies with respect to the different possible situations have to be found and identified. Therefore, it is rather an issue of identifying the most effective measures and actions first and not of being efficient in the execution of the actions. According to [7], the proof of effectiveness for automated driving functions can be quite tricky. Anyhow, the scenario database in combination with relevance measures from the test field is a very powerful method for the efficient development and validation of automated driving and traffic control functions. In fact, "effectiveness" is a validation criterion by itself.

4.5.6 Cosimulation and Virtual Sensors

Not all necessary control parameters can be measured directly. For example, danger, collision risk, congestion probability, capacity reserve, etc., are values which cannot be measured or evaluated directly during a test run. Either these are measured with so-called virtual sensors, or cosimulations are executed with the test runs, where the desired values are taken from the simulation results then. The scenario database can also be seen as a form of cosimulation, only that the simulations are executed in advance and special methods help to access the data in time. That way, the scenario database can be a much easier variant of cosimulation. Traditional cosimulation may technically be very tricky, though. Virtual sensors can be implemented easily with the help of machine learning, where said unmeasurable criteria are trained from the simulations in the scenario database. Afterwards, they can be applied on the sensor measurements of the test run, enabling to have the "unmeasurable" values available in real time during the test runs.

4.5.7   Complexity and Robustness Management

When doing automation and control (traffic as well as vehicle control), this is mainly about optimisation with respect to certain control targets. When optimising, it must be talked about robustness as well in general. Squeezing a system(-of-systems) to be optimal, this normally tends to lack robustness. Therefore, robustness must be assessed compulsory. One way for doing such is the described scenario-based approach. Within the scenario catalogue, not only different alternative control actions are evaluated. In a similar way also, perturbations and disturbances are evaluated to find out if the control solutions tend to become disastrous in case of small errors. Such robustness must be incorporated in any design and therefore must be part of any validation procedure. While robustness normally can only be checked in post-analysis, there is a way to improve robustness from the beginning, by reduction and control of complexity. Measuring complexity of traffic flows, behavioural patterns, data streams, control solutions, infrastructure, etc., can help to reduce complexity by design. Reduced complexity makes life easier and more robust for human traffic as well as for automated traffic.

References

1. Hons C, Neubohn A, Weiss C, Keck F, Kuhn A (2007) Kiss-weiterentwicklung und umsetzung eines methodischen ansatzes zur systemauslegung in der fahrzeugsicherheit. VDI-Berichte (2012)
2. Keck F, Kuhn A, Sigl S, Altenbuchner M, Palau T, Roth F, Stoll J, Zobel R, Kohsiek A, Zander A (2010) Pruef-und evaluationsverfahren fuer den vorausschauenden fussgaengerschutz im spannungsfeld zwischen simulation und realer erprobung. VDI-Berichte (2106)
3. Kuhn A, Carmona J, Novak T, Aigner W, Schildorfer W, Patz D (2018) Test fields and advanced accompanying methods as necessity for the validation of automated driving. In: Proceedings of 7th transport research arena TRA 2018, Vienna, Austria
4. Kuhn A, Carmona J, Palau T (2015) A hierarchical, subsidiary system architecture for traffic control with connected vehicles. Technical report
5. Kuhn A, Eibl G, Fasig T (2012) Concept for an "intelligent" traffic control network. In: 19th ITS world congress ERTICO-ITS European Commission ITS America ITS Asia-Pacific
6. Neubohn A, Weiss C, Keck F, Kuhn A (2005) Kiss—A universal approach to the development and design of occupant restraint systems. In: Proceedings: international technical conference on the enhanced safety of vehicles, vol 2005. National Highway Traffic Safety Administration, pp 13–13

7. Sigl S, Gollewski T, Miehling T, Kuhn A (2014) About development processes and accompanying performance evaluations of integral automotive safety systems. In: Proceedings of FISITA 2014 world automotive congress, Maastricht

Andreas Kuhn studied Technical Mathematics and Mechanical Engineering at the Vienna University of Technology. There he also awarded his Ph.D. for the simulation of special satellite dynamics. He now works for more than two decades in several positions and roles in the fields of automotive safety, automated driving and traffic automation with an steady focus on virtual development procedures and the safe application of softcomputing methods.

José Carmona Ph.D. in Computer Science and Computer Science Engineer from the University of Málaga (Spain), is a development engineer at Andata GmbH. His research interest include computational intelligence, data mining and simulation in connection with traffic automation.

Carmona Elvira Thonhofer has received her Masters Degree in Mechanical Engineering from the Vienna University of Technology. Her research interests include traffic modelling, simulation and control, intelligent transportation systems and automation. Elvira has worked on national research projects and her academic contributions are published in relevant journals and conferences.

Part II
Assessment Methodologies and Their Application

After setting the scene for truck platooning in Part I, truck platooning assessment methodologies and their application in the Connecting Austria project are detailed in Part II. Due to the heterogeneous and multidisciplinary research endeavour, research results relevant for different research domains are summarised in this part. The results cover investigations of aerodynamic drag effects due to reduced inter-vehicle distances and their validation on test tracks. Furthermore, study results related to comprehensive investigations of truck platoon dynamics and traffic effects are presented in Part II. Thereby, also the facilitation of truck platooning through intelligent road infrastructures, which are able to monitor traffic and provide relevant information to other road participants, is investigated. Finally, Part II presents an approach to assess truck platoon efficiency and its application within the Connecting Austria project.

Chapter 5
Truck Platoon Slipstream Effects Assessment

Alexander Kospach and Christoph Irrenfried

Abstract With the increase of stringent emission standards and higher road transportation cycles in the last few decades, the importance of transport and fuel efficiency plays a major role. The aerodynamic forces on trucks have a huge impact of the overall fuel consumption rate. For a 40 tonnes semi-trailer truck at 85 km/h on a flat highway, around 40% of the provided engine power is needed to overcome the air resistance (Hucho in Aerodynamik des Automobils. Vieweg + Teubner, Wiesbaden, [1]). An efficient way to reduce the aerodynamic drag of trucks is to build a platoon of trucks. To assess the potential of a truck platoon due to slipstream effect, computational fluid dynamic (CFD) simulations were conducted. The simulations were performed for a platoon with three trucks for different constant velocities at different inter-vehicle distances. The results are summarised in a normalised drag coefficient and fuel reduction map. As a limiting factor of platooning, the thermal management aspect must be considered, because the slipstream reduces the air mass flow through the engine compartment. This aspect of reduced air mass flow through the engine compartment was analysed as well.

5.1 Computational Setup

In this section, the computational setup is described which was used for simulations.

The original version of this chapter was revised by including a new coauthor. The correction to this chapter can be found at https://doi.org/10.1007/978-3-030-88682-0_18

A. Kospach (✉)
Virtual Vehicle Research GmbH, Graz, Austria
e-mail: alexander.kospach@v2c2.at

C. Irrenfried
Graz University of Technology, Institute of Fluid Mechanics and Heat Transfer, Graz, Austria

© The Author(s) 2022, corrected publication 2022
A. Schirrer et al. (eds.), *Energy-Efficient and Semi-automated Truck Platooning*,
Lecture Notes in Intelligent Transportation and Infrastructure,
https://doi.org/10.1007/978-3-030-88682-0_5

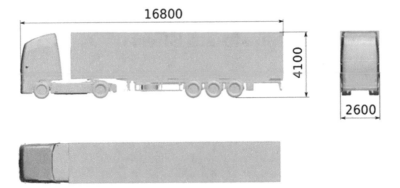

Fig. 5.1 Used truck model with the outer dimensions in mm

5.1.1 Model Geometry and Virtual Wind Tunnel

A simplified semi-trailer truck model was used for simulations and is shown in Fig. 5.1 with its outer dimensions in mm.

Major simplifications were made for the wheel rims. Also, detailed components (e.g. the antenna, cabin suspension, foot board) were not considered because the influence of this components on the investigated aerodynamic parameters is negligible.

A virtual wind tunnel was created for the CFD investigations and is shown in Fig. 5.2. The box around the trucks consists of the boundaries inlet, outlet, floor and symmetric walls. The different boundary conditions for these boundaries will be explained below in Sect. 5.1.2. The outer dimensions were chosen in such a way that the influence of the truck geometry on the boundary condition is insignificantly small. The trucks are arranged in the middle of the wind tunnel and the front edge of the first truck is 50 m away from the inlet boundary. In Fig. 5.2, the flow direction and the outer dimensions of the wind tunnel are shown.

5.1.2 Boundary Conditions

The boundary conditions for the simulations, as illustrated in Fig. 5.2, are:

- **Inlet:** At the inlet boundary condition, a constant velocity u_∞ is specified, based on the considered truck speed.
- **Outlet:** In order to guarantee a defined pressure level, the pressure at the outlet boundary is kept constant.
- **Symmetric walls:** In order to guarantee a gradient-free velocity field at the boundary surfaces, a symmetrical boundary condition is prescribed at these surfaces.
- **Wheels:** The wheels are modelled with rotating wall boundary conditions.

- **Floor:** The floor boundary condition is modelled with a moving wall boundary condition with a constant velocity u_∞.
- **Truck surfaces:** All truck surfaces except the wheels are assigned a no-slip boundary condition.

5.1.3 Heat Exchanger Model

The engine compartment flow is considered in the simulation. In the engine compartment, three heat exchangers are modelled, namely the water cooler, condenser and charge air cooler. In Fig. 5.3, the position of the heat exchangers is shown, condenser (green), charge air cooler (blue), water cooler (red). All three heat exchangers are modelled as porous media using the Darcy–Forchheimer law [6]. The Darcy–Forchheimer law describes the relationship between pressure loss and flow velocity based on the Darcy coefficient D and Forchheimer coefficient F.

The coefficients can be experimentally identified by measuring the pressure loss of the individual heat exchangers at different volume flows. Using the method of least squares, the coefficients D and F can be determined from measurements. Table 5.1 shows the values used in the simulation based on experimental data:

5.1.4 Mesh Generation for Simulation

The mesh generation for simulation was done in OpenFOAM with the mesher "snappyHexMesh". The mesher "snappyHexMesh" creates, based on the assigned model,

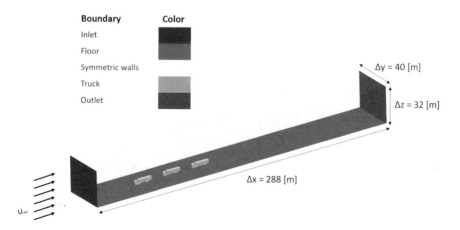

Fig. 5.2 Virtual wind tunnel with its outer dimensions and the used boundaries

Table 5.1 Experimentally identified coefficients of the Darcy–Forchheimer model

Engine part	F (1/m)	D (1/m^2)
Charge air cooler	36	4.20×10^6
Water cooler	320	5.97×10^7
Condenser	320	1.4×10^7

a hexahedral dominated mesh. The base cell size was defined as 2 m, with the refinement boxes placed in close proximity to the truck. Figure 5.4 illustrates the arrangement of these refinement boxes. The different refinement levels define the mesh resolution in the refinement boxes based on the base level with the relation

$$\text{cell size} = \frac{\text{base size}}{2^{\text{refinement level}}}. \tag{5.1}$$

The truck geometry had a maximal refinement level of 8 on the walls. Inside the engine compartment, the refinement of critical geometry areas went up to level 11. The resulting mesh is shown in Fig. 5.5 as a longitudinal section (cut through $y = 0$ m), for a single truck case. It can be seen that with higher levels, the cell sizes are

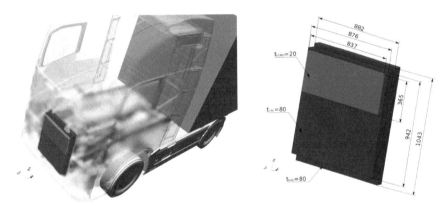

Fig. 5.3 Position of the heat exchangers in the engine compartment and their dimensions, specified in mm

Fig. 5.4 Arrangement of the refinement boxes for meshing

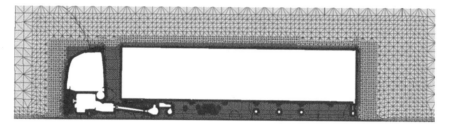

Fig. 5.5 Simulation mesh for the longitudinal section

Table 5.2 Material
parameters for the simulation
model

Parameter	Value
ρ_{air}	1.2043 kg/m^3
ν_{air}	1.5126×10^{-5} m^2/s
p_{ref}	101,325 Pa
T_{ref}	20 °C

reduced. For the solo truck case, the mesh consisted of around 22 million cells, and
for the platooning cases with 3 trucks, the simulation mesh consisted of around 66
million cells.

5.1.5 Flow Field Computation

The CFD simulations were performed with the GNU GPL licensed software pack-
age OpenFOAM. For the investigated aerodynamic problem, the assumption of a
incompressible, steady-state flow with constant material properties were made and
using an Reynolds-averaged Navier–Stokes equations (or RANS equations) approach
[5]. The underlying RANS conservation equations were solved with the OpenFoam
"SimpleFoam" solver. SimpleFoam solver is a steady-state solver for incompressible, turbulent flow, using the Semi-Implicit Method for Pressure Linked Equations
(SEMI) algorithm [4]. As turbulence model, the k-ω-SST model was chosen, which
has already proven its accuracy and efficiency in automotive aerodynamics [3]. The
k-ω-SST model is a two-equation eddy-viscosity model for the turbulence kinetic
energy, k, and turbulence specific dissipation rate, ω. The used material property
values for the density ρ_{air}, the kinematic viscosity ν_{air} and the reference pressure p_{ref}
and temperature T_{ref} are shown in Table 5.2.

5.2 Simulation Results and Discussion

The simulations were done for a platoon with three trucks for different constant velocities at different inter-vehicle distances between each truck. The inter-vehicle distances within the platoon were kept constant for each simulation. The investigated velocities were 60, 80 and 90 km/h. For all velocities, the distances 6.66, 11.11, 22.22, 33.33, 56.56 m were simulated, and for the 80 km/h, the additional spacing of 15 m was simulated.

5.2.1 Drag Coefficients

The main key parameter to quantify the drag or the resistance of an object is the drag coefficient c_d. For all simulations, the drag coefficient was evaluated and compared. The drag coefficient is defined as

$$c_d = \frac{2F_d}{\rho u^2 A} \tag{5.2}$$

where

- F_d is the drag force, which is by definition the force component in the direction of the flow velocity,
- ρ is the mass density of the fluid,
- u is the flow speed of the object relative to the fluid,
- A is the reference area [7].

For automobiles and many other objects, the reference area is the projected frontal area of the vehicle [1]. The drag coefficients for the different investigated velocities were nearby the same. The relative deviation for the drag coefficients for the velocities 60 and 90 km/h compared to drag coefficients for the velocity 80 km/h was less than 2%. Thus, the drag coefficient can be assumed to be independent of the driving speed for the investigated velocities, and just, the results for the velocity 80 km/h are shown.

In Fig. 5.6, the normalised drag coefficients are shown for the velocity 80 km/h. The drag coefficient was normalised by the drag coefficient for the solo truck simulation. The greatest savings potentials are at the smallest inter-vehicle distance. Even truck 1, which only shows a change in the trailing area behind the truck, shows a reduction of about 25% of the resistance value of the solo truck case. For the smallest distances 6.66 and 11.11 m, truck 2 has a smaller drag coefficient than truck 3. It can be seen that for larger distances (15, 22.22, 33.33 and 56.56 m) truck 3 has the smallest drag coefficient followed by truck 2 and truck 1 through the slipstream effect. Truck 1 and truck 2 except for the 15 m case have a rising trend of the normalised drag coefficients. For truck 3, there is a decreasing trend for smaller distances, and from 22.22 m, there is an increasing trend. A decreasing of the averaged drag coefficient of the platoon could be observed for all simulations for a decreasing inter-vehicle distance except for 15 m. On average, up to 30% reduction of the drag coefficient is possible by using a platoon.

In terms of flow analysis, two major factors are responsible for the drag reduction on the vehicles in the platoon. These are lower incoming flow velocity which reduces the pressure on the front and higher pressure on the back.

Figure 5.7 shows the kinematic pressure distribution of all cases for the longitudinal section. The kinematic pressure here is defined as followed using the static pressure p_{stat}, the reference pressure p_{ref} and the density ρ_{air}:

$$p_{kin} = \frac{p_{stat} - p_{ref}}{\rho_{air}}. \tag{5.3}$$

The effect of a higher pressure on the back is clearly seen for smaller distances. The higher pressure on the back is the reason why for the smallest distances truck 2 has a smaller drag coefficient than truck 3 and why there is a significant drag reduction for the leading truck 1. The effect of a higher pressure on the back is also observed at large distances (33.33 and 56.56 m) for truck 1 resulting in slightly reduced drag coefficients compared to the solo case. Figure 5.8 shows the velocity distribution for the longitudinal section for all cases. The lower incoming flow velocity for the following trucks 2 and 3 can be seen, which results in reduced pressure in the front. This can also be seen in Fig. 6.11 of Chap. 6, where the pressure drag coefficients c_p are shown for front and back for truck 2 for selected cases.

In Fig. 5.9, the cumulative normalised pressure drag coefficients for the cases 6, 11, 15 and 22 m for the second truck are shown to get a hint how the drag coefficients occur. The cumulative pressure drag coefficients are plotted using 20 subdivision

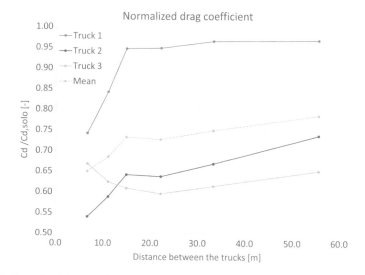

Fig. 5.6 Normalised drag coefficients for the velocity 80 km/h

with a spacing of around 84 cm. It can be seen that for the 15 m case the initial cumulative normalised pressure drag coefficient is higher than for the other shown cases, which leads to an overall higher drag coefficient. Higher maximal stagnation pressures in the middle of the front of truck 2 for case 22 m can be observed compared to case 15 m but for case 15 m higher pressures at the side of the front areas, roof of the cabin and the cab-roof fairing are obtained. For the other cases, the expected increasing trend of the cumulative normalised pressure drag coefficient can be seen. Also, the effect of a higher pressure on the back can be observed resulting in a lower total pressure drag coefficient, especially for case 6 m.

5.2.2 Fuel Savings

In order to be able to conclude from the reduction of the drag coefficient to a reduction of fuel consumption, the total share of drag losses in the total mechanical energy must be determined. It can be assumed that the relative percentage of engine losses in relation to the total energy from the fuel remains the same. The internal engine losses do not change due to the changed air resistance. Therefore, in order to evaluate the savings potential through the reduction of air resistance, the relative proportion of air resistance to the required mechanical energy must be considered [1]. Thus, the reduction of the drag coefficient with a form factor M, using

$$\Delta c_d = M \cdot \Delta T, \tag{5.4}$$

can be converted to the fuel saving ΔT (in litre/100 km) [2]. A typical value of $M = 3.5$ is chosen for these studies, given in [2]. The potential fuel savings are

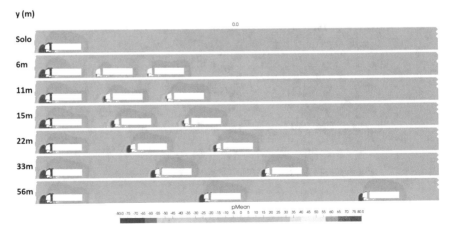

Fig. 5.7 Kinematic pressure distributions for the longitudinal section for all cases

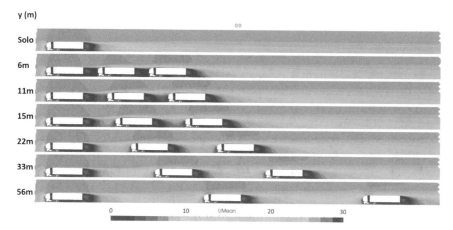

Fig. 5.8 Velocity distributions for the longitudinal section for all cases

shown in Fig. 5.10 using the simulation results for the reduction of the drag coefficient from Fig. 5.6. On average, a fuel reduction up to 10% is realised for the smallest investigated spacing.

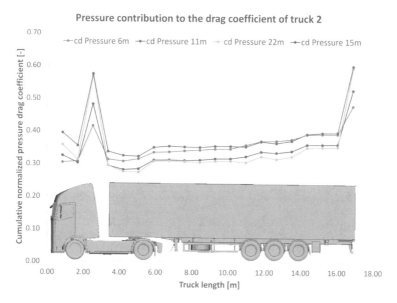

Fig. 5.9 Cumulative normalised pressure drag coefficient for the inter-vehicle distances 6, 11, 15, 22 m for the second truck

5.2.3   Mass Flow Through Heat Exchangers

As seen in the previous results, the platooning effect can lead to significant reduction of the pressure in the front and the air mass flow through the heat exchangers is reduced, which leads to a reduced cooling performance. For all three heat exchangers, the volumetric mass flow was evaluated to be able to identify potential risks concerning thermal management issues. The volumetric mass flows were taken and normalised with the volumetric mass flow of the solo truck case. In Fig. 5.11, the normalised mass flows over the radiator are shown. It can be seen that there is a significant mass flow reduction over the heat exchangers for truck 2 and truck 3, especially for the smaller distances. The mass flow for the charge air cooler and condenser is not shown because they have similar results like the shown radiator case. For the case with an inter-vehicle distance of 6 m, the air mass flow drops under 50% for truck 2 and truck 3. This low air mass flows imply a risk for the thermal management system. These implications on the thermal management system are targeted in Chap. 15. This is important because the additional needed fan power for small distances increases the fuel consumption and the gains from platooning drag reduction can be diminished.

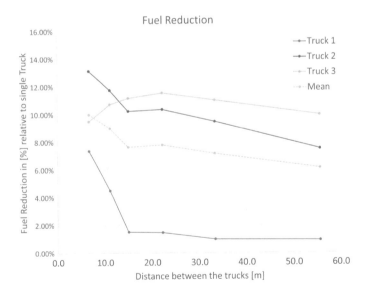

Fig. 5.10 Fuel reduction in % relative to single truck

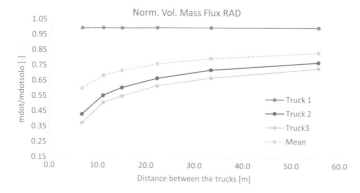

Fig. 5.11 Normalised volumetric mass flow over radiator for a varied inter-vehicle distance

5.3   Conclusion

Computational fluid dynamics (CFD) simulations are conducted for platoons of three vehicles. Several inter-vehicle distances and velocities were investigated. Further, it is apparent that there is a significant potential in fuel savings for smaller distances. It can be seen that there are also critical inter-vehicle distances like 15 m where the drag reduction is not following a decreasing trend to smaller distances for truck 2. Further investigations are needed to be able to better delimit the critical distance range. The results concerning the mass flow over the heat exchangers show a potential risk for the thermal management for the smallest distances. This issue is addressed in Chap. 15. Further simulative investigations with different model geometries and with different arrangements would be beneficial to assess their influence on the total drag reduction.

References

1. Hucho WH (2005) Aerodynamik des Automobils. Vieweg + Teubner, Wiesbaden
2. Indinger T, Devesa A (2012) Verbrauchsreduktion bei nutzfahrzeug-kombinationen durch aero-dynamische massnahmen. ATZ - Automobiltechnische Zeitschrift 114(7):628–634
3. Menter FR, Kuntz M, Langtry R (2003) Ten years of industrial experience with the SST turbu-lence model. Turbulence Heat Mass Transfer 4(1):625–632
4. Moukalled F, Mangani L, Darwish M (2016) The finite volume method in computational fluid dynamics : an advanced introduction with OpenFOAM® and Matlab, volu 113 of Fluid mechan-ics and its applications. Springer, Cham
5. Spurk J, Aksel N (2010) Strömungslehre - Einführung in die Theorie der Strömungen. Springer, Berlin, Heidelberg, New York
6. Whitaker S (1996) The forchheimer equation: a theoretical development. Transport Porous Media 25(1):27–61
7. Wikipedia Contributors (2003) Drag coefficient. Accessed 21 Jan 2003

Alexander Kospach is Team Leader at the Virtual Vehicle Research GmbH for "Advanced Measurement and Fluid Flow Simulation". Together with his team he is involved, also as coordinator, in several international and national research projects concerning heat transfer (battery and power electronic cooling, cabin comfort), aerodynamic optimisation, gas measurement technology and synchronised wireless information platform DATA.BEAM.

Christoph Irrenfried is an Assistant Professor and head of the aerodynamics group of the Institute of Fluid Mechanics and Heat Transfer at Graz University of Technology, Austria. He received his Ph.D. from the same university for research on convective turbulent heat transfer in wall-bounded flow, using direct numerical simulations in combination with experimental validation. From 2017 to 2019, he was a Senior Researcher at the Virtual Vehicle Research GmbH. His research interests include aerodynamics, experimental and computational fluid dynamics.

Chapter 6
Validation of Truck Platoon Slipstream Effects

Bernhard Lechner, Almir Cajic, Bernhard Fischbacher, Alexander Kospach, Alexander Mladek, Peter Sammer, Christoph Zitz, Michael Zotz, and Christoph Irrenfried

Abstract Due to slipstream effects, platooning leads to a significant decrease of the fuel consumption of the heavy-duty vehicles (HDV). Measurements with a platoon consisting of three vehicles were performed at the Zalazone proving ground. The goal of these measurements was to get the static pressure at the front and the rear of the second vehicle to calibrate computational fluid dynamics simulation and to measure the fuel consumption directly. Measurements were done at a vehicle speed of 80 km/h and varying inter-vehicle distances. Platooning leads to a reduction of the pressure coefficients in the centre of the HDV front and an increase of the pressure coefficient at the top and the rear of the HDV. Furthermore, a reduction of the fuel consumption of the leading vehicle of 7.9% at an inter-vehicle distance of 6 m and 3.7% at a distance of 22 m was determined. A comparison to CFD simulation showed a similar fuel reduction for an inter-vehicle distance of 6 m and 22 m. CFD simulation showed an increase of fuel consumption at an inter-vehicle distance of 15 m. This increase was experimentally not validated. Also, results for the following vehicle are presented.

Keywords Air Flow · Fuel consumption · Heat rejection · Measurement · Platooning · Pressure drop · Proving ground · Slipstream · Vehicle distance · Wireless

The original version of this chapter was revised by including a new coauthor. The correction to this chapter can be found at https://doi.org/10.1007/978-3-030-88682-0_18

B. Lechner (✉) · A. Cajic · B. Fischbacher · A. Kospach · A. Mladek · P. Sammer · C. Zitz · M. Zotz
Virtual Vehicle Research GmbH, Graz, Austria
e-mail: bernhard.lechner@v2c2.at

C. Irrenfried
Graz University of Technology, Institute of Fluid Mechanics and Heat Transfer, Graz, Austria

© The Author(s) 2022, corrected publication 2022
A. Schirrer et al. (eds.), *Energy-Efficient and Semi-automated Truck Platooning*,
Lecture Notes in Intelligent Transportation and Infrastructure,
https://doi.org/10.1007/978-3-030-88682-0_6

6.1 Introduction

Platooning is defined as electronically coupling a group of vehicles at close inter-vehicle distances to either increase the capacity of the road or reduce fuel consumption due to slipstream effects. It is estimated that fuel consumption and exhaust gas emissions can be reduced up to 10% by platooning [1–4]. The share of fuel cost on the total costs in the transportation business is 21.2% [5]. The earnings of truckage companies are usually very low. Between 2015 and 2017, the earnings were between 1.8 and 3.0% [6]. Thus, even a small reduction of fuel consumption can lead to significant increase in earnings. There are three approaches to estimate the potential of reducing fuel consumption by platooning:

1. **Real-world measurement**—For real-world measurements, heavy-duty vehicles (HDV) are equipped with a limited measurement system, especially a GPS and a system for acquiring data of the on-board diagnostic system (OBD). Thus, long-term measurements are possible. However, it is very difficult to synchronise data of different HDVs of the platoon, it is impossible to get ambient parameters, and above all, there are strict legal restriction to the minimum inter-vehicle distance.
2. **Computational fluid dynamics simulation (CFD)**—Although the possibilities of computational fluid dynamics increased significantly in recent years, CFD has to rely on calibration data provided by measurements.
3. **Measurements on a proving ground**—For measurements on a proving ground, the platoon's vehicle can be fully equipped, ambient conditions (wind speed and wind direction e.g.) are well known, measurements can be repeated several times, and it is possible to investigate very low inter-vehicle distances. In this work, the minimum distance was chosen to be 6 m. The mayor drawback of proving ground measurements is the limited length of the test section. In order to increase the aerodynamic effects, the vehicles have to go as fast as possible. At a test section of 400 m and a vehicle speed of 20 m/s, there are just 20 s to align the vehicles and conduct the measurements.

In the following, measurements on the Zalazone Proving Ground [7] with a platoon consisting of three HDVs are described. The main goal was to provide data of the static pressure at the vehicle surface to validate the CFD simulation and to quantify the influence of the inter-vehicle distance on the fuel consumption.

6.2 Materials and Methods

The proving ground, the vehicles and the sensor set-up are described in this section. Also, the mathematical models to calculate pressure coefficient and fuel consumption are presented.

Fig. 6.1 Test section at the
Zalazone Proving Ground,
base map and map data from
OpenStreetMap,
© OpenStreetMap
contributors under the
CC-BY-SA license, https://
www.openstreetmap.org/
copyright

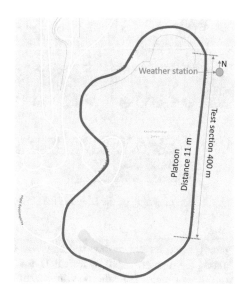

6.2.1 Proving Ground

The measurement campaigns were done at the handling course of the Zalazone Proving Ground (see Fig. 6.1). This handling course has a total length of 2200 m and the straight section of 400 m, which was used as a test section for platooning measurements. It has a slight slope of approximately 1 m. This elevation has to be taken into account for the pressure measurements. A platoon with three vehicles of a length of 17 m and a distance of 11 m is illustrated for clarity in Fig. 6.1.

6.2.2 Heavy-Duty Vehicles

The platoon consisted of three heavy-duty vehicles (HDV) of type Volvo FH 540. The trailer of the HDV was empty. The weight of the HDV was approximately 17,000 kg. Due to this lightweight, the engines of the HDV were operated at an engine load of 25%.

6.2.3 Sensors

Data Acquisition System

The HDVs of the platoon were equipped with data acquisition systems which had to be synchronised below millisecond level (see Fig. 6.2). At all HDVs, a data acquiring

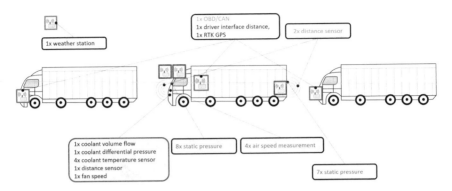

Fig. 6.2 Sensor system on the HDVs of the platoon

system based on the on-board diagnostic system (OBD) and a GPS was installed. Among the data provided by the OBD, fuel rate, engine speed and engine load were used for the analysis of the fuel consumption. The second HDV of the platoon was additionally equipped with the following measurement systems:

1. Eight sensors were measuring the static pressure at the front of the HDV, and four Prandtl tubes were installed at hood air intakes.
2. In order to monitor the engine's cooling systems, the radiator was equipped with a coolant volume flow sensor, a differential pressure sensor and a temperature difference sensors.
3. The static pressure at the vehicle rear was measured at seven locations.
4. The distance to the preceding HDV was measured by a laser distance sensor. A similar system was installed at the third HDV.

Pressure Sensors

Eight flat pressure probes (Type FDStat60 by SVMTec) were mounted at the front and seven at the rear of the vehicle. These probes were connected to the pressure sensors (Honeywell, 006MDSA3, range ± 600 Pa) by small tubes. However, these sensors are differential pressure sensors. Thus, a reference pressure is needed. Clearly, this reference pressure can neither be taken from ambient (as the HDVs are moving) or from the drivers cabin as an underpressure is induced by the passing air flow. Therefore, the pressure sensors were connected to a thermally isolated reference pressure vessel. In Fig. 6.3, the positions of the pressure probes are indicated.

Ambient

Speed and direction of the ambient air were measured using a 3D ultrasonic anemometer. As can be seen in Fig. 6.1, the weather station was located at the end of the test section. As wind speed and direction can change along the test section, this anemometer only gives information regarding the overall wind speed and direction during the measurements.

Fig. 6.3 Flat pressure sensors mounted on the front (left) and rear of the second HDV (right)

6.2.4 *Measurement Campaigns*

Two measurement campaigns were conducted at Zalazone Proving Ground in July and August 2019. The focus of the first campaign was on the pressure measurements and on the fuel consumption during the second campaign. All measurements were done at a target vehicle speed of 80 km/h or 22.2 m/s. For each campaign, the first measurements were done with a single reference vehicle. Afterwards, the measurements were done with the platoon at different distances. At the first campaign, the inter-vehicle distances were set to 6, 11, 15, 22 and 55 m. For the second campaign, the distances were 7, 15 and 22 m. The measurements were repeated several times, usually five times during the first campaign and nine times during the second campaign.

6.2.5 *Static Pressure*

The static pressure $p_{i,\text{stat}}$ at position (x_i, y_i) of the ith sensor position on the HDV is the sum of the dynamic pressure and the barometric pressure, given by

$$p_{i,\text{stat}} = \rho\, c_{\text{p},i}\,(x_i, y_i, d_{\text{front}}, d_{\text{back}}, \Delta y_{\text{front}}, \Delta y_{\text{back}})\, \frac{v^2 + w^2 + 2vw\cos(\beta)}{2} \qquad (6.1)$$
$$+\, a_h \Delta h.$$

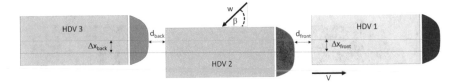

Fig. 6.4 Factors influencing static pressure and fuel consumption

The dynamic pressure (first summand) depends on the ground speed v, the wind speed w, the direction β of the wind speed with respect to the driving direction as well as the coefficient of pressure $c_{p,i}$, see Fig. 6.4. The conversion of the dynamic pressure to the static pressure and thus also $c_{p,i}$ is influenced by:

1. **Geometry**—The conversion depends strongly on the position of the sensor. In the centre of the HDV's front (e.g. sensor 4 and sensor 6 in Fig. 6.3), nearly all the dynamic pressures are converted to static pressure, while at the roof, even a negative static pressure can be induced (e.g. sensor 8 in Fig. 6.3).
2. **Distances**—The air flow can be blocked by the preceding or following vehicle. The distances to the preceding vehicle d_{front} and following vehicle d_{back} influence the static pressure and therefore the fuel consumption.
3. **Lateral offset**—If one vehicle has a lateral offset to another, the air flow is only partially blocked. Thus, Δx_{front} and Δx_{back} also influence static pressure, drag and fuel consumption.

The barometric pressure $a_h \Delta h$ [second summand in (6.1)] decreases by 12 Pa per metre and can be seen well in the measurement of the sensors at the rear of the HDV.

6.2.6 Data Preprocessing

While GPS data and OBD data were sampled at 50 Hz, other parameters were measured at a slower rate (e.g. pressure at 2 Hz). Thus, in a first step, all measurement data were interpolated to the same sampling rate. Acceleration was not measured directly, but calculated from the GPS speed. As the involved derivative calculation significantly increased the noise level, acceleration was smoothed by a Savitzky–Golay filter. Data was sampled continuously during the whole lap. The data sampled at the test section was extracted for further analysis. In most of the following charts, the data of the test section is indicated by red shading.

6.3 Results

The goal of this measurement campaign was to provide data to validate the computational fluid dynamics (CFD) simulation of the platoon and to directly measure fuel consumption and the influence of inter-vehicle distance.

6.3.1 Static Pressure

In a first step, the measurements were done with a reference vehicle to get the static pressure without influence of the platoon. Afterwards, measurement with a platoon consisting of three HDVs was done with different inter-vehicle distances.

Reference Vehicle

The static pressures measured at eight positions in the front of the reference HDV (see Fig. 6.3) are shown in Fig. 6.5. The data of the whole measurement is shown. The vehicle drove three laps. The data of the test section is indicated by red shading and can be easily identified due to the approximately constant pressure profile. The mean values of the sensors at the test section are also given in Fig. 6.5. As the vehicle speed was 21.6 m/s and the wind speed was 0.8 m/s coming from south east, the dynamic pressure is calculated to 273 Pa. The sensors $p_{\_stat\_4}$ and $p_{\_stat\_6}$ are in the centre of the enginehood, and c_p is close to one. Sensor $p_{\_stat\_1}$ located in front of the licence plate also got a c_p close to unity. Contrary, at sensor $p_{\_stat\_8}$, even negative pressure values were measured. The pressure at the laterally positioned sensors $p_{\_stat\_2}$ and $p_{\_stat\_3}$ and $p_{\_stat\_5}$ and $p_{\_stat\_7}$ varies between 115 and 240 Pa.

At the test section, the static pressure values measured at the rear of the HDV varied between -11 and -28 Pa (Fig. 6.6). At all sensors, the static pressure declined by approximately 10 Pa. This is due to the elevation of the test section by one metre.

Model of the Static Pressure at the Centre of the Front of the Reference Vehicle

For the reference vehicle, Eq. 6.1 simplifies significantly. As d_{front} and d_{back} as well as Δx_{front} and Δx_{back} vanish, c_p is constant. Thus, $c_{p,i}$ and a_h can be determined by linear regression, since the static pressure as well as the vehicle speed, the wind speed, the wind direction and height are measured. The results for sensor $p_{\_stat\_4}$ are

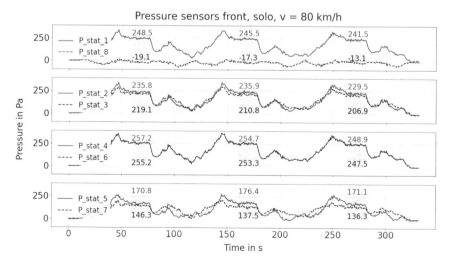

Fig. 6.5 Static pressures at the front of the reference HDV

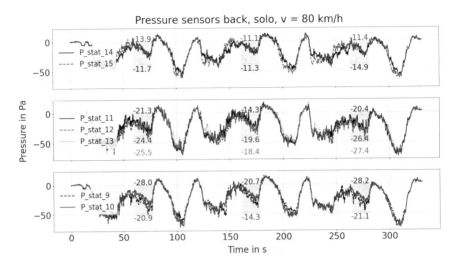

Fig. 6.6 Static pressures at the rear of the reference HDV

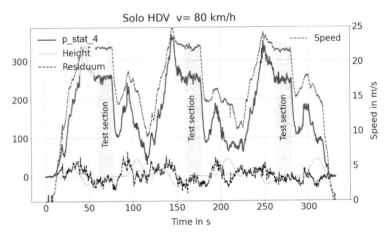

Fig. 6.7 Static pressures at position 4, vehicle speed, height and residuum of the model

shown in Fig. 6.7 as an example. Data of all other sensors behaves the same way. The adjusted coefficient of determination is 0.99, indicating that the independent data can very well described by the model of Eq. 6.1. The independent variables speed and height are shown as well as the dependent variable p_{stat_4}. Again, the range of the test section is highlighted by a red background. The residuum of the linear regression is also shown in Fig. 6.7. The model explains the measurement data very well for the test section, where the model overestimates the static pressure of sensor 4 only by approximately 4%. In Fig. 6.8, the residuum along the handling course is shown. Length and colour of the arrows indicate the normalised residuum, while the direction of the arrows corresponds to the velocity vector. It can be seen

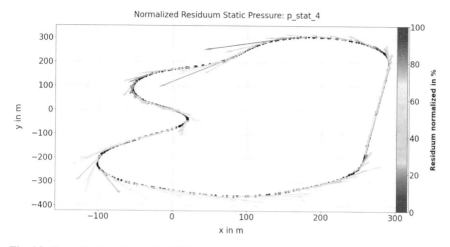

Fig. 6.8 Normalised residuum (0–100%) and velocity vector along the handling course

that the residuum is very low at the test section and largest when the vehicle travelled south-west. In Table 6.1, the pressure coefficients are given for the the eight sensors at the front and for the seven sensors at the back of the reference vehicle. In the columns "TS LAP 1", "TS LAP 2" and "TS LAP 3", the pressure coefficient is calculated with the data of the test section for each lap, only. In the column "Total", data of the whole handling course was used to calculate the pressure coefficient. There is a variation of 5% for the test section measurement. The pressure coefficient for the whole handling course is higher. As was shown in Fig. 6.8, c_p is overestimated when the vehicle goes south-west.

Pressure Coefficients for the Platoon

The static pressure for the sensors mounted on the second HDV of the platoon is shown in Fig. 6.9. The vehicle speed was set to 22.2 m/s. However, there was a tailwind of 2 m/s coming from south/southeast. The platoon went for four laps. Again, the data of the test section is highlighted in red. Comparing Figs. 6.5 and 6.9, the flow profile differs significantly. The pressure values of sensor 1, mounted in the middle of the license plate, are reduced from 250 Pa for the reference vehicle to even negative values of lap 4 for the following vehicle. On the other hand, values at sensor 8 near the roof increase from −20 to 70 Pa. At the first lap, a strong decrease of the measured pressure can be seen. For this measurement, the drivers did not succeed in aligning the HDV but reduced the distance constantly. During the first lap, the mean distance between the first and the second HDV was 9.6 m, between the second and the third vehicle 15.4 m. The pressure coefficient calculated by linear regression for the second vehicle of the platoon (front as well as rear) is given in Table 6.2. The coefficient of determination calculated by linear regression is 0.934, so somewhat smaller than for the reference vehicle, but nevertheless indicating that 93.4% of the variance can be explained by this model. c_p increases with the inter-vehicle distance. However, even at a distance of 55 m, c_p is significantly lower than at the reference vehicle.

Table 6.1 Pressure coefficient of the reference HDV

| Sensor (front) | Pressure coefficient $c_{p,i}$ | | | |
	Total	TS Lap 1	TS Lap 2	TS Lap 3
$p_{\_stat\_1}$	0.99	1.00	0.992	0.976
$p_{\_stat\_2}$	0.946	0.956	0.954	0.929
$p_{\_stat\_3}$	0.861	0.889	0.855	0.839
$p_{\_stat\_4}$	1.03	1.04	1.03	1.01
$p_{\_stat\_5}$	0.704	0.696	0.718	0.698
$p_{\_stat\_6}$	1.02	1.03	1.02	1.00
$p_{\_stat\_7}$	0.60	0.60	0.564	0.56
$p_{\_stat\_8}$	−0.044	−0.055	−0.047	0.03
Sensor (rear)	Total	TS Lap 1	TS Lap 2	TS Lap 3
$p_{\_stat\_9}$	−0.08	−0.089	−0.059	−0.09
$p_{\_stat\_10}$	−0.017	−0.061	−0.074	−0.061
$p_{\_stat\_11}$	−0.052	−0.062	−0.084	−0.072
$p_{\_stat\_12}$	−0.072	−0.075	−0.0578	−0.084
$p_{\_stat\_13}$	−0.072	−0.079	−0.051	−0.086
$p_{\_stat\_14}$	−0.029	−0.025	−0.024	−0.037
$p_{\_stat\_15}$	−0.026	−0.032	−0.023	−0.025

Table 6.2 Pressure coefficient at the front and rear of the second HDV of the platoon for a varied inter-vehicle distance

| Sensor (front) | Pressure coefficients $c_{p,i}$ | | | | |
	6 m	11 m	15 m	22 m	55 m
$p_{\_stat\_1}$	0.218	0.294	0.452	0.476	0.678
$p_{\_stat\_2}$	0.180	0.335	0.478	0.367	0.640
$p_{\_stat\_3}$	0.294	0.382	0.396	0.490	0.579
$p_{\_stat\_4}$	0.209	0.380	0.493	0.482	0.708
$p_{\_stat\_5}$	0.369	0.352	0.559	0.343	0.609
$p_{\_stat\_6}$	0.221	0.400	0.507	0.486	0.717
$p_{\_stat\_7}$	0.357	0.409	0.400	0.427	0.402
$p_{\_stat\_8}$	0.171	−0.035	0.001	−0.015	0.048
Sensor (rear)	6 m	11 m	15 m	22 m	55 m
$p_{\_stat\_9}$	0.061	0.014	0.007	−0.027	0.014
$p_{\_stat\_10}$	0.060	0.015	0.015	−0.024	0.030
$p_{\_stat\_11}$	0.061	0.011	0.01	−0.03	0.021
$p_{\_stat\_12}$	0.049	0.017	0.09	−0.027	0.020
$p_{\_stat\_13}$	0.051	0.01	0.09	−0.021	0.015
$p_{\_stat\_14}$	0.054	0.032	0.02 1	−0.002	0.040
$p_{\_stat\_15}$	0.058	0.026	0.022	−0.007	0.039

There is a significant increase of the pressure coefficient at inter-vehicle distance of 15 m. All sensors beside p_stat_3, p_stat_7 and p_stat_8 show significant higher pressure values. Interestingly, also CFD simulation shows a significant increase of the pressure coefficient at a inter-vehicle distance of 15 m.

6.3.2 Fuel Consumption

Reference Vehicle

Acceleration and deceleration are the most fuel-consuming driving patterns. The reference vehicle used cruise control which minimises the number of acceleration and deceleration events. Additionally, it is not necessary to adjust distance and lateral offset to the other vehicles. Therefore, fuel consumption of the reference could be measured accurately. At an engine speed of 1340 RPM, the average fuel consumption at the test section was measured to 5.81 ± 0.26 ml/s. However, measurement were done at different engine speeds. Using data from the first vehicle of the platoon, a correction factor was calculated:

$$f_{\text{rate}} = 0.0027 N_{\text{eng}} + 1.68 \tag{6.2}$$

Thus, for the reference vehicle, at an engine speed of 1750 RPM, a fuel rate of $f_{\text{rate}} = 6.8$ ml/s is given.

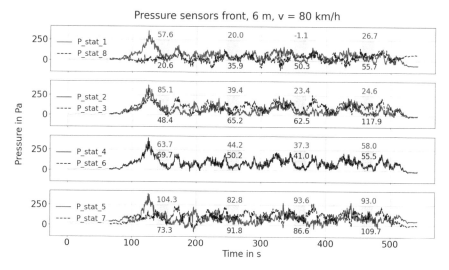

Fig. 6.9 Static pressures of the sensors in front of the second HDV. Distance to the preceding HDV 6 m

First Platooning Vehicle

As for the reference vehicle, the first vehicle does not have to accelerate and decelerate, and the fuel consumption is not strongly superimposed by acceleration effects. It is assumed that the fuel consumption is influenced by acceleration, engine speed, distance and offset of the following HDV. Vehicle speed has a big influence to the fuel consumption. However, at this measurement set-up, vehicle speed was constant at 22.2 m/s. As least square regression can only minimise variance (and there was none), vehicle speed was not included in the model. Additionally, there is a strong correlation between vehicle speed and engine speed. Including both variables to the model resulted in colinearity problems. The model used for estimating the fuel rate f_{rate} is:

$$f_{rate} = a_0 a + a_1 N_{eng} + a_2 \Delta X_{1-2} + a_3 \Delta Y_{1-2} \tag{6.3}$$

with the acceleration a in m/s^2, the rotational speed of the engine N_{eng} in revolutions per minute/1000, the distance between the first and the second HDV ΔX_{1-2} in m and the lateral offset of the second vehicle ΔY_{1-2} in metres. Other models were tested but did not produce statistically significant results. The parameters a_i were estimated by linear regression and are shown in Table 6.3. The most influential parameter is acceleration. The fuel rate increases by 12.05 (ml/s)/(m/s^2). The fuel rate due to engine speed increases by 3.57 (ml/s)/(RPM/1000). The influence of the distance is described by a_2. It is small, however statistically significant. The fuel rate increases by 0.018 (ml/s)/m. The influence of the lateral shift of the preceding vehicle is even smaller. The fuel rate increases by 0.0041 (ml/s)/m.

Table 6.3 Regression parameter for the fuel rate for the first vehicle of the platoon

Parameter	Expected value	Lower limit	Upper limit
a_0	12.05	11.59	12.52
a_1	3.57	3.56	3.59
a_2	0.018	0.016	0.02
a_3	0.0041	0.0039	0.005

Table 6.4 Regression parameter for the fuel rate for the second vehicle of the platoon

Parameter	Expected value	Lower limit	Upper limit
a_0	24.24	23.79	24.69
a_1	3.59	3.52	3.67
a_2	0.0083	0.001	0.02
a_3	0.0007	−0.003	0.004

Table 6.5 Fuel savings due to platooning for a varied inter-vehicle distance

Fuel savings

Vehicle	Expected value			Lower limit			Upper Limit		
	6 m	15 m	22 m	6 m	15 m	22 m	6 m	15 m	22 m
HDV_1 (%)	7.9	5.6	3.7	7.2	4.6	2.5	8.3	6.2	4.6
HDV_2 (%)	8.2	7.14	6.3	5.2	2.6	0.54	10.6	10.5	10.4%

Following Vehicles Within the Platoon

The same model (6.3) was applied to the second vehicle of the platoon. The regression parameter is shown in Table 6.4. Interestingly, the influence of the acceleration is twice as high as for the first vehicle and the influence of the engine speed is the same as for the first vehicle. The parameter a_2 describing the influence of the distance is smaller. The lateral offset to the first vehicle is statistically not significant. The fuel saving of the platoon compared to a single HDV is shown in Table 6.5. For the leading HDV, the fuel saving is between 7.9% at a distance of 6 m and 3.7% at 22 m. The upper and lower limits of the confidence interval are also given in Table 6.5. The fuel saving of the second HDV is 8.2% at a distance of 6 m, 7.14% at 15 m and 5.2% at 5.2 m. However, the error of margin is much higher than for the first HDV.

6.3.3 Comparison to Simulation Results

A comparison of the fuel consumption of the CFD results with measurement data was done. As shown before, the direct fuel consumption values can only be used for the first truck. In comparison with the following trucks, only the first truck has an evenly sufficient drive (use of the cruise control), which means that the measured reduction in fuel consumption can be clearly attributed to the reduction in air resistance. A comparison of the consumption reduction of this measurement with the CFD simulations is shown in Fig. 6.10. For the smallest distance 6 m, a very good agreement between measurement and simulation is obtained. A deviation of 4–5% can be observed for the distance 15 m. Also, the effect that the fuel reduction of distance 15 m is nearby the same as for distance 22m could not be seen for the measured fuel data but can be seen in the derived pressure drag coefficients for front and back from measured data, see Fig. 6.11. As explained in this chapter the measured fuel rate data was quite sensitive to external factors. For the 22 m point, the absolute error is smaller than for the 15 m point and is around 2%. It can also be observed that the measured values are always greater than the simulated values.

Another comparison of the simulation results was done using the pressure signals from the front and back of truck 2. The exact measurements of these pressure signals and position of the sensors is shown in this chapter. The pressure signals where also evaluated for the simulation results. The same position of the pressure sensors like

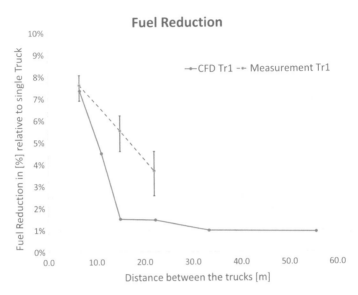

Fig. 6.10 Comparison of simulation and measurement results of the fuel reduction in [%] relative to single truck

in the measurements was used. Some pressure measurement points are located in areas with local pressure gradients. In order to determine the influence of the local sensor position on the pressure results, the static pressure values of the simulations on the surface of the front and rear sides were averaged for different radii. A displacement of the sensor up to 3 cm has no noticeable influence on the result. Since the positioning accuracy of the pressure sensors is within this range, the influence of the positioning can be neglected for all further evaluations. The pressure values were used to calculated an arithmetic average pressure value for the front and back sides. The average pressure value is used to calculate the pressure drag coefficient. For calculation of the pressure drag coefficient, the stagnation pressure and reference pressure were used. The drag coefficient can be divided into two components, namely frictional drag coefficient (viscous drag) and pressure drag coefficient (form drag), whereas for our cases, the frictional drag coefficients are insignificantly small in relation to pressure drag coefficients. In Fig. 6.11, the different pressure drag coefficients for front and back are shown. The solo case with a theoretical distance of infinity is represented here with a distance of 100 m because of illustration. The pressure drag coefficient values for the front have a good agreement for all compared cases except at the point 15 m. Here, the measured value is higher. For the pressure drag coefficient values in the back deviations for all distances can be observed. The measured pressure drag coefficients at the back are greater than the pressure drag coefficients obtained from simulations, which indicates that the simulations might overpredict the negative pressure and underestimate the influence of the trailing truck on the pressure distribution at the back. It seems the deviation of the measured and

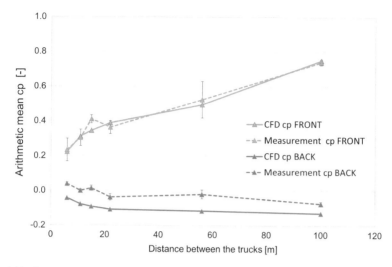

Fig. 6.11 Comparison of simulation and measurement results for the pressure coefficients front and back for truck 2

simulated pressure drag coefficients at the back has a nearby constant offset for all reliable data points and the decreasing trend of the pressure drag coefficient at the rear side of the truck is covered by simulation.

6.4 Discussion

The measurement of the fuel consumption of a platoon compared to a single HDV poses a great challenge to the instrumentation of the HDV, the measurement campaign and the data processing.

6.4.1 Instrumentation

While the instrumentation of an HDV poses a challenge for itself, the sticking point of the instrumentation is the highly synchronised data acquisition. Due to the high speed of the HDV of 22 m/s, even a time shift of ten milliseconds results in an error of the distance measurement of 22 cm. An overview of all sensors can be seen in Fig. 6.12.

The measurement system "DATA.BEAM" is developed at the Virtual Vehicle Research GmbH for distributed measurements with precision time synchronisation. During measurements, all "DATA.BEAM" devices on the trucks are synchronised via 868 MHz wireless communication to an average error of 10 μs. Immediately

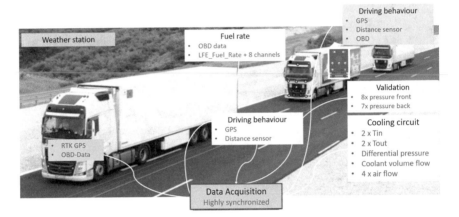

Fig. 6.12 Platoon with a sketch of all sensors

after the measurement, "DATA.BEAM" collects the measurement data. Using this procedure, it is possible to substantially speed up the measurements and reduce the effort of post-processing as all data are already synchronised.

6.4.2 Measurement Campaign

There are two possibilities measuring fuel consumption of a platoon vehicle: either a measurement campaign at a test section, as it was done in this project, or real world on road measurements. Both approaches have advantages and disadvantages. The measurement campaign at the Zalazone Proving Ground has the main advantage that important parameters like wind speed and wind direction can be measured and the measurement can be repeated under the same boundary conditions. The major drawback was the very short test section of only 400 m. Thus, the HDV passed the test section in less than 20 s. During this short period, the HDV had to be aligned and the distance had to be adjusted. It was very difficult to reach a steady state of the platoon without any acceleration events. This can be seen in Fig. 6.15. The positions of the HDV of the platoon are shown every two seconds. The platoon has been shifted along the y-axis for better visibility. The platoon entered the test section and was aligned after 6 s. After 8 s HDV 3 was too close. The distance to HDV 2 was 3.5 m instead of 7 m. This was corrected by the driver. However, this manoeuvre required some deceleration. The platoon was stable after 12 seconds. After 14 seconds HDV 2 was too close to HDV 1. After 18 seconds, the first HDV reached the end of the test section and the platoon was dissolved.

To reduce the effect of acceleration, data with an acceleration above a threshold of 0.2 m/s^2 was discarded. In Fig. 6.13, an histogram of the acceleration at the test

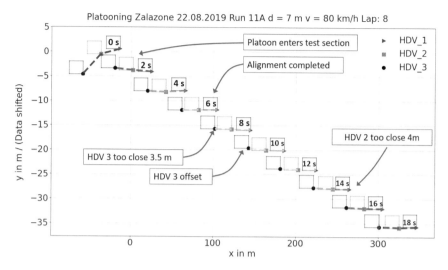

Fig. 6.13 Alignment of the HDV at the test section

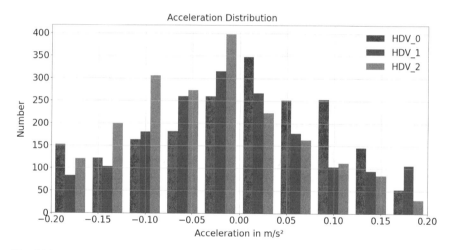

Fig. 6.14 Acceleration events at the test section

section is shown. According to Tables 6.3 and 6.4, even an acceleration of 0.05 m/s^2 leads to an additional fuel rate of 0.5 ml/s for the trailing and 1 ml/s for the following HDV. The fuel rate versus engine speed is shown for a distance of the HDV of 7 m for all laps. The black lines indicate the maximum fuel rate at 100% engine load. The fuel rate for different laps is indicated by a different colours and marker symbols. It can be seen that due to the empty trailer, the measurements took place at low fuel rates. As described, all data above an absolute value of 0.2 m/s^2 have been discarded. Nevertheless, the fuel rate varies between 1 and 15 ml/s. The analysis of the least square model suggests that approximately 50% of this variation can be explained by

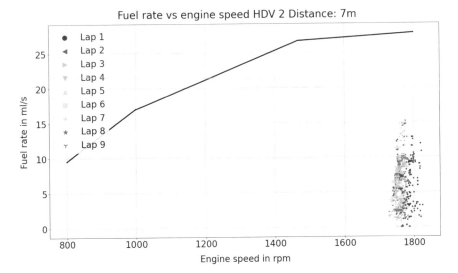

Fig. 6.15 Fuel rate over engine speed for HDV 2 at a distance of 7 m

the acceleration. Figure 6.14 an histogram of the acceleration at the test section is shown.

6.4.3 Lessons Learned

The measurement campaigns revealed that highly accurate and synchronised data of a moving platoon consisting of three HDV can be provided. Especially, due to the measurement procedure provided by "DATA.BEAM", it was possible to do the measurement according to a very strict time table. As the measurement data was automatically synchronised and put into data groups, post-processing could be reduced to a minimum. However, it became evident that a test section of 400 m is too short. At a inter-vehicle distance of 22 m, the length of the platoon is already 95 m; thus, a quarter of the test section is occupied by the platoon itself. It is therefore highly recommended to use a test section with a minimum length of 800 m.

References

1. Al Alam A, Gattami A, Johansson KH (2010) An experimental study on the fuel reduction potential of heavy duty vehicle platooning. In: 2010 13th International IEEE annual conference on intelligent transportation systems, pp 306–311
2. Dhamale BP, Jadhav NP (2017) Aerodynamic simulation of a truck using platooning technique. Int Eng Res J

3. Lammert M, Duran A, Diez J, Burton K (2014) Effect of platooning on fuel consumption of class 8 vehicles over a range of speeds, following distances and mass. SAE In J Commer Veh 7(2):738–743
4. Liang K-Y, Martensson J, Johansson KH (2013) When is it fuel efficient for a heavy duty vehicle to catch up with a platoon. In: 7th IFAC symposium on advances in automotive control, pp 738–743 (2013)
5. Damiaa P, Jorge D-L, Javier B (2019) Estimating road transport costs between EU regions. Technical report. JRC—Joint Research Center
6. Naderer Ruth (2019) Branchenreport Speditionen und Logistik 2019. Technical report. Kammer für Arbeiter und Angestellte Wien
7. Automotive Proving Ground Zala Ldt. Handling ground, H-1055 Budapest, Honvéd u., pp 13–15

Bernhard Lechner Study of Physics at the Karl-Franzens-University of Graz, Ph.D. at the Institute of Internal Combustion Engines and Thermodynamics of the University of Technology, Graz about exhaust emissions of aircraft engines. At the Virtual Vehicle Research GmbH, responsible for advanced measurement systems with a focus on exhaust gas aftertreatment and thermal investigation of radiators.

Almir Cajic is a master's degree student in Mechanical Engineering at the Graz University of Technology. His specialization subjects are "Automotive Engineering and Safety" and "Computational Engineering and Mechatronics". Almir is a parttime employee at the Virtual Vehicle Research GmbH where he is involved in research projects in the area of measuring technology for the automotive industry.

Bernhard Fischbacher finished his diploma degree in Telematics with emphasis on embedded automotive engineering in the year 2015 at GrazUniversity of Technology. Since 2012 he works at Virtual Vehicle Research GmbH in the group of "Advanced Measurement and Fluid Flow Simulation", facing diverse measurement and data acquisition tasks both as project member and project manager. Bernhard is currently working on all topics related to emission measurement as well as supporting the planning and execution of measurement campaigns.

Alexander Kospach is Team Leader at the Virtual Vehicle Research GmbH for "Advanced Measurement and Fluid Flow Simulation". Together with his team he is involved, also as coordinator, in several international and national research projects concerning heat transfer (battery and power electronic cooling, cabin comfort), aerodynamic optimisation, gas measurement technology and synchronised wireless information platform DATA.BEAM.

Alexander Mladek holds a master's degree in Aeronautical Engineering. He is working at Virtual Vehicle Research GmbH as a Senior Researcher in the field of Thermal Management and Mobile Air Conditioning HVAC. His research interests cover efficient thermal management and comfort systems as well as machine learning applications. He is involved in international and national research projects with his special focus on software integration and thermal simulation.

Peter Sammer is senior researcher at Virtual Vehicle Research GmbH in the area of advanced measurement and fluid flow simulation. He studied telematics and information and computer engineering at the Technical University Graz. His area of expertise ranges from hard- and software development in distributed applications to algorithm design for automated driving. The data acquisition platform "DATA.BEAM" is the main focus of his work where he developed an algorithm for wireless time synchronisation in sensor networks.

Christoph Zitz has studied mechanical engineering and business economics with focus on energy technology on the Technical University of Graz. He has been working on the Virtual Vehicle since 2012 as a student employee and from February 2018 as a junior researcher in the group "Advanced Measurement and Fluid Flow Simulation". His main tasks are the construction of test benches, the support of the cooler test bench and the implementation of aerodynamic measurement tasks in the in-house wind tunnel at Virtual Vehicle Research GmbH.

Michael Zotz is a student at the Universitiy of Technology in Graz. Beside his studies of mechanical engineering he works for Virtual Virtual Vehicle Research GmbH and is part of the work group "Advanced Measurement and Fluid Flow Simulation". Dealing with CAD for design as well as manufacturing and assembling the test rigs is covered by his tasks. In addition he supports measurement tasks about the wind tunnel and individual test beds.

Christoph Irrenfried is an Assistant Professor and head of the aerodynamics group of the Institute of Fluid Mechanics and Heat Transfer at Graz University of Technology, Austria. He received his Ph.D. from the same university for research on convective turbulent heat transfer in wall-bounded flow, using direct numerical simulations in combination with experimental validation. From 2017 to 2019, he was a Senior Researcher at the Virtual Vehicle Research GmbH. His research interests include aerodynamics, experimental and computational fluid dynamics.

Chapter 7
Simulation of Platoon Dynamics, Optimisation and Traffic Effects

Elvira Thonhofer and José Carmona

Abstract This chapter outlines the methodologies required to realise a comprehensive scenario-based approach for effective and efficient development and validation of complex, cooperative control functions in connected and automated driving. These methods are exemplified for platooning and are devised in the scope of Connecting Austria, the Austrian flagship project on automated driving and goods transport. The development and validation approach have first been implemented vertically in depths for the intersection use cases of Connecting Austria. The scenario-based approach includes

- The systematic identification, collection and collocation of the relevant and representative traffic scenarios.
- The modelling and simulation of the according traffic and vehicle control strategies.
- The effectiveness assessment of the traffic and vehicle control strategies with the help of suitable key performance indicators.
- The controlled iterative adaption to new situations and boundary conditions by steady extension of the operational design domain within an adaptive, learning framework.

The demonstration use case "intersection" is the most complex with respect to possible C-ITS, traffic and vehicle control actions. That way generality should be guaranteed, enabling a quick, horizontal extension to further use cases and scenarios, aiming to cover all relevant situations for platooning vehicles within their operational design domain. The application of all methods introduced here will be demonstrated in Chap. 9.

Keywords Effectiveness assesment · Scenario management · Scenario-based function development · Cooperative connected automated driving · Platooning

E. Thonhofer (✉) · J. Carmona
Andata Artificial Intelligence Labs, Hallein, Austria
e-mail: elvira.thonhofer@andata.at

© The Author(s) 2022
A. Schirrer et al. (eds.), *Energy-Efficient and Semi-automated Truck Platooning*,
Lecture Notes in Intelligent Transportation and Infrastructure,
https://doi.org/10.1007/978-3-030-88682-0_7

89

7.1 Methodology for Scenario-Based Analysis

The scenario-based approach mentioned in Chap. 4 is carried out comprehensively for the realistic and relevant situations of demonstration use case "intersection" in the small town of Hallein in the region of Salzburg in Austria. The investigated area comprises several complex controlled intersections in a sequence with some roundabouts, reaching from the highway exit through the city to some logistic root and end nodes. All these traffic nodes contain mixed traffic under complex topological conditions with various conflicting situations of individual vehicles, trucks, bicycles, pedestrians, powered two-wheelers and public transport. The given traffic conditions are complex enough to inherit several critical issues and circumstances, which may lead to complicated collapses of traffic in the area if interest.

7.1.1 Traffic Detection

First, one needs a clear understanding and overview about the real traffic situations and scenarios in detail. Three different approaches are used and combined here. In the first instance, a system with video cameras is installed for traffic detection in the selected area of interest. The cameras detect, classify and make anonymous all traffic participants by deep learning methods and precisely evaluate their trajectories as illustrated in Fig. 7.1. A detailed analysis of the traffic situation in Hallein, Austria, is shown in Chap. 10.

Hereby, the trajectories of all traffic participants are evaluated and merged consistently for all the consecutive traffic nodes of the controlled intersections and roundabouts. These trajectories are later used for the detection, interpretation and prediction of the traffic situations.

In addition to the video cameras, also lidar sensors are used for traffic detection. Such measurements can be seen in Fig. 7.2.

In comparison with the video detection, the lidar sensors are more robust with respect to light and weather conditions. Object detection and classification are simpler than with videos from the algorithmic point of view. Making data anonymous is not an issue here, because objects cannot be detected at a personal level. On the other hand, these sensors are much more expensive than simple video cameras. Hence, the sensors are used for detailed analysis of selected street sections only.

A further source for detecting and classifying the traffic situations is floating car data, like they are provided from different vendors and sources. Figure 7.3 illustrates the application of machine learning-based clustering algorithms for the identification of the relevant traffic situations. Hereby snapshots from floating car data are taken periodically over a relevant time range and area of interest. The floating car data is then clustered and classified with unsupervised machine learning algorithms. The shown map is generated automatically and summarises all traffic situations according to the representation of the floating car data, which really occurred in the time and

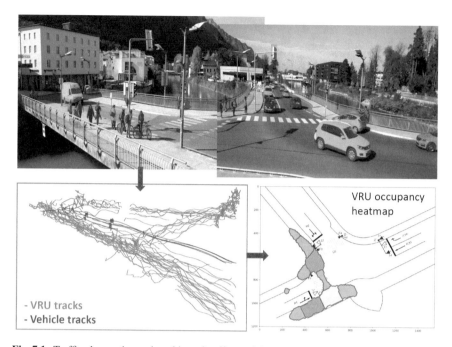

Fig. 7.1 Traffic observation and tracking of traffic participants with video and deep learning algorithms

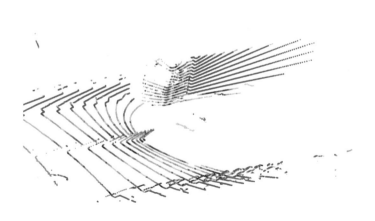

Fig. 7.2 Traffic detection with lidar sensors

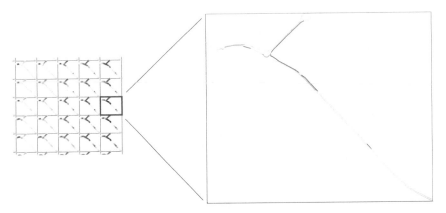

Fig. 7.3 Detection and classification of traffic situations by floating car data: similar situations are grouped in clusters in the catalogue of traffic situations (*left*)

region of investigation. Beside the comprehensive identification and representation of the really occurring traffic situations, these are not only clustered and listed but also arranged with respect to similarity. Similar traffic situations are neighbours in the shown catalogue of traffic situations.

7.1.2 Naturalistic Driving and Field Operational Tests

The traffic detection described in Sect. 7.1.1 is accompanied by naturalistic driving data, delivering additional data for the validation of the trajectories from the vehicle's internal view, tracking some data from the vehicle bus and further sensors for the driver behaviours and vehicle surroundings. While the data from Sect. 7.1.1 corresponds to the Eulerian specification from the perspective of flow dynamics, the data evaluated here in Sect. 7.1.2 corresponds to the Lagrangian specification.

Lagrangian data is primarily used for the development of proper behavioural model of drivers and arbitrary traffic participants, which requires that external objects must be identifiable by the vehicle sensors. One of the behavioural aspects that can be modelled and calibrated by using the naturalistic driving data is reaction times and interactive behavioural patterns. For example, when some vehicles are waiting at a traffic light, the first vehicle does typically not start moving immediately when the traffic light changes to green, but only after a certain reaction time, and the second vehicle does not start moving as soon as the vehicle in front initiates the movement. As a consequence, delay times accumulate, slightly reducing intersection capacities. By analysing the underlying naturalistic driving data, it is possible to model such reaction times in a stochastic way and use these models inside simulation models for improved accuracy and realism.

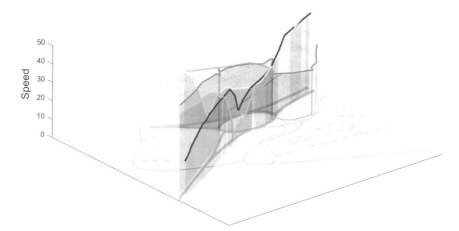

Fig. 7.4 Vehicle tracks from naturalistic driving studies

Another relevant aspect is the vehicle following behaviour, which describes the reaction in terms of speed variation of a vehicle in lateral direction with respect to the movement of the vehicles ahead. It influences the distances between vehicles depending on the speed and driver characteristics. Following behaviours have a noticeable impact on the capacity of the intersection, since they affect the maximum density of traffic in the street which is still sufficiently safe.

A further aspect puts focus on the street sections and statistical variations of driving behaviours with respect to the spatial context. Figure 7.4 shows examples of three individual vehicle tracks from different drives in the investigated area. One easily can see the different speed profiles, the vehicles where driving in the given section around the selected roundabout. The speed profiles of course correspond to the topology of the street sections and the spatial surroundings, the traffic situation and driver characteristics. In the investigated area, significantly more than 1000 individual trips with different drivers driving two specially equipped cars have been used for the calibration of stochastic driver models.

7.1.3   Traffic Modelling

The traffic situations as well as the proper models for driver behaviours are incorporated into agent-based micro-simulation models for the whole area under investigation. In Sect. 9.1.1, such a model of a specific intersection in Hallein is introduced. The traffic models are calibrated and validated intensively and carefully with the data from the traffic detection in Sect. 7.1.1 as well as the single vehicle tracks from the naturalistic driving studies from Sect. 7.1.2. Special care is taken in the consideration of interactive behaviours of the different traffic participants. The traffic light signal coordination plans are also included in the model, along with all relevant traffic signs and priority rules.

As part of the validation, the key performance indicators that result from the simulations, such as travel times, vehicle counts and others, are compared with the various sensors measurements. The traffic model is then steadily and iteratively refined until no significant discrepancy between simulation and reality remains. The performance indicators must be analysed in a stochastic way, since micro-simulation traffic models (same as traffic in general) are not deterministic. It is furthermore necessary to vary the initial and boundary conditions of the simulation. As a result, a massive database of simulation results is made available, which offers a very detailed understanding of how different control strategies and actions affect the performance indicators under different traffic situations, including numerical conflicts between different key performance indicators. This is further described in Sect. 7.1.4.

One further advantage of the comprehensively validated micro-simulation traffic models is that the whole detailed information of each agent (cars as well as vulnerable road users) is available, including positions, speed and many other features for each time stamp. That makes it possible to find out which data aggregation and filtering mechanisms work better for each control strategy. Additionally, it allows to evaluate how reliable the aggregated floating car data is depending on how many vehicles provide their information to the data aggregation service. Confidence intervals for the aggregated floating car data can also be calibrated by analysing micro-simulation results.

7.1.4 *Development of Functions by Scenario Management*

Having valid numerical models for arbitrary conditions is a core element of the scenario-based development and validation procedure. Once having valid and trust-worthy simulation models, not only the given control strategies but also various alternative *vehicle and traffic control strategies* and actions are carried out in the simulations.

Namely numerous different variants of the scenarios are produced with the help of Monte Carlo simulation approaches. Different aspects are subject to such variations. An obvious example is traffic situations, which can be described using the inflows to the system, the routing decisions or origin–destination matrix and the vehicle composition. Pedestrian flows are of course an important component of the traffic situation. Environmental conditions can be varied too. The interactive behaviours of traffic participants are also subject to variations, including reaction times, longitudinal behaviour and car-following models, lane changing behaviour, overtaking and many others. Dynamic components of the infrastructure are also varied, including traffic light signal plans and V2I/I2V messages. Even static components of the infrastructure can be subject to variation, including aspects such as the number of lanes, new road markings and others.

These massive variations are executed and simulated on top of existing reference control algorithms for vehicle and traffic control. Traffic regulations and laws are considered as being static. In principle, these could be changed in a similar way, too. The reference controls are traditional static traffic control coordination plans in combination with manually driven vehicles under consideration of their driving behaviours, taken from Sect. 7.1.2.

A comprehensive set of relevant fundamental manoeuvres is identified, thoroughly studied and automated via trajectory optimisation and platoon control techniques. One of those is the individual extension of green times for the platoon in order to allow platoons to pass the intersection completely and not being split by an unlucky green time phasing. Coordinated drive-away at the start of the green phase has also been investigated and assessed as a control strategy able to minimise the accumulated reaction time discussed in Sect. 7.1.2. A further strategy which has been evaluated is the temporal and spatial increase of platooning densities for an optimised passage of a given fixed green time phase. Such a strategy allows platoons to pass the intersection without splitting and to improve the capacity of the intersection. Other control strategies that have been evaluated are the energy optimised approach of intersection at red light phase as well as the prioritisation of certain traffic participants (like public transport or emergency vehicles).

These control actions required implementing the relevant aspects of V2X communication into the simulation models, too. In the case of platooning, such control strategies may include the selection of situation-aware distance gaps, for example. Other strategies include using the I2V information from the traffic lights about remaining time until the next green or red phase to optimise the speeds and spacing of the platoon members, or, inversely, using V2I information from the platoon to adapt the signal plan in order to give priority to the platoon, if so desired.

Safely managing cut-in-, cut-through- and unexpected braking manoeuvres have been implemented and tested in vehicle dynamics simulations, too. The aspects of safe automated driving have been addressed by state-of-the-art and novel scientific results including safety-extended predictive platoon control and measures to obtain string stability, an essential dynamic property in efficient and safe automated driving systems.

7.1.5 Evaluation and Analysis of Key Performance Indicators (KPIs)

For the simulations of the various control strategies and actions, the resulting key performance indicators (KPIs) are specified, evaluated and assessed. These do not only include traditional indicators like travel time losses, as shown in Fig. 7.5, flow capacities, etc., but also more sophisticated parameters like capacity reserves, collision risks, congestion risks, complexity of the traffic situation, robustness, duration of the traffic disturbance caused by an event, such as a platoon being given some additional green time to cross the intersection, etc.

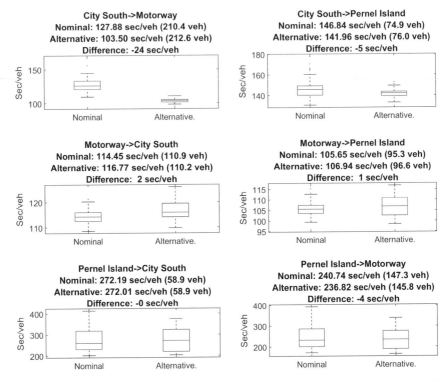

Fig. 7.5 Exemplary evaluation and comparison of travel and loss times for certain routes with reference solution and an alternative control strategy in the investigated area of interest

Some of these KPIs can be derived and measured directly from single simulations or measurements. Others, like collision risks and danger, congestion risk, flow capacity reserves, criticality rates and some others cannot be measured directly, neither in simulation nor in the real world. For such KPIs, virtual sensors are used, which have been developed and applied with special multi-layered stochastic simulation methods. Their description exceeds the scope of this exposition by far and therefore will be disclosed by subsequent publications.

Robustness is another important property for the rating of control algorithms, which needs to be inherited in system and control algorithm design imminently. For example, the traffic loss time of the nominal reference solution significantly scatters due to the natural variations. These can be seen in the box plots of Fig. 7.5, which show the comparison of travel times for the reference solution on the left and for an alternative, improved control solution on the right. Beside the reduction of travel times, also the scatter of the travel times could be reduced significantly, expressing increased robustness of the system performances.

Complexity is another relevant property of traffic situations. The novel concept of traffic complexity rating has been first introduced in [1]. It allows to make decisions such as not allowing specific control strategies, when the resulting traffic situation becomes too complex. In general, complexity correlates with the loss of resilience of the traffic control system and can be used as a design criterion to avoid non-robust control strategies. A high traffic complexity rating is a good indicator for congestion risk.

The corresponding information design helps in a concerted selection of the best and situation-aware control strategies and actions. There are typically conflicts between different key performance indicators. Therefore, it is necessary to focus on all the relevant and available indicators at once. A typical example is that a longer green phase for one direction improves the traffic flow in that direction but makes it worse in the perpendicular direction at a signalised crossing.

Figure 7.6 graphically summarises the effect of a given control strategy on the whole system, coding the resulting aggregated speeds for each street section using a colour code, like it is known from various floating car data representations.

It has to be noted that multiple complementary simulation environments with varying granularity, simulation scope and depth ranging from traffic flow micro-simulations to detail vehicle control and vehicle dynamics simulations are utilised to cover the multiple aspects and scopes of the posed research questions, that is, the different performance indicators. For example, a research question on potential benefits and risks of utilising platoon control to realise coordinated passing of a signalled intersection needs to be answered broadly:

- Traffic simulations allow statements on the traffic flow effects in the surrounding traffic network links.
- Detailed platoon vehicle control/vehicle dynamics simulations are needed to assess the response characteristics of such manoeuvres in the presence of mixed/individual traffic and to verify the behaviour of situation-aware platooning control.
- Measurement data and derived behavioural models must be analysed to obtain valid model parameters and assumptions for any simulation study.

Finally, dedicated tests on real roads—from testing grounds to open-road tests are needed to back up, extend or challenge the findings and to allow iteration loops to refine research questions, tools and methods, as well as data support and close the research feedback loop.

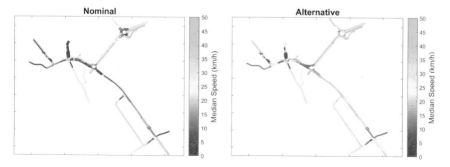

Fig. 7.6 Comparison of resulting network traffic performance as a result of applying alternative control strategies (right)

7.1.6 Adaptation and Learning

One of the reasons for the complexity and the difficulties of traffic control and management is the fact that human traffic participants are adapting to different traffic situations and traffic control strategies dynamically and sometimes unexpectedly. In such cases, the capability of quick adaption to new traffic situations is the only available resolution. Therefore, after application and implementation of the improved traffic and vehicle controls, their effect is not only evaluated in simulations under given, certain assumptions. The control strategies and their implementations must be assessed in the real environment as soon as possible, according to Sect. 7.1.1. Changes in traffic participants' behaviours or the change in the underlying assumptions due to new control strategies are captured hereby quickly and re-integrated in the scenario catalogue, enabling a controlled, iterative enhancement and improvement for the steady adaption to new situations and the realisation of cooperative learning. This way, the cycle for continuous development and (self-)adaptive learning of complex system-of-systems environments is closed. Of course, this implies that the real environment must be monitored continuously in order to systematically identify, describe and track such changes in traffic situations and behaviours of the traffic participants. One of the methodological fundamentals for such a continuous iterative process is anomaly and incident detection. Detecting anomalous, not previously observed situations, allows the developers to identify, model and simulate them in order to evaluate the effects of the different strategies and actions in these newly observed situations. This way, anomaly detection guarantees that all relevant situations are identified, collected and collocated in the scenario catalogue, and that changes in the underlying real-world situations are incorporated systematically in the scenario catalogue. Anomaly detection for new traffic situations can be carried out using simple statistical techniques, but also complex and powerful machine learning methods that can recognise anomalous traffic patterns.

7.2 Integral Safety and Advanced Driver Assistance Systems (ISS/ADAS)

Recalling Fig. 4.1 in Chap. 4, safety constitutes a major benefit of automated driving in general and of platooning in particular. A portion of the research effort therefore is concentrated on automotive safety.

Our approach is founded on the following principles:

- Assuming a probabilistic/stochastic point of view.
- Consequent top-down instead of bottom-up system requirements development.
- Analysis of field effectiveness instead of test effectiveness.
- Increasing integration of simulation.

7.2.1   Use Case-Based Representation of Requirements

Adopting use case-based specifications can greatly simplify the task of defining required system behaviour by allowing the system engineer to focus on one use case at a time. The system engineer can then assign the desired system behaviour to each use case. Of course the number of use cases required to sufficiently specify the system depends on the problem to be solved. Covering the complete field of operation of ISS/ADAS may easily require several thousand use cases (note that we use this term in a broad sense that includes "misuse cases" as well). In the scope of the project Connecting Austria, for example, platoon safety is of particular interest. The relevant use case is collision avoidance while platooning. Details regarding the application of this method are demonstrated in Chap. 9.

Stochastic simulation allows to generate the required diversity of use cases in an economically feasible way. However, this raises the need for new methods to cope with large numbers of requirements, including the following:

- Heuristics or physically motivated rules must be derived that allow an automatic assignment of the desired system behaviour to each use case.
- Numerical conflict analysis techniques can identify requirement conflicts (or prove their absence) in a large set of use cases.
- Metrics are needed to monitor the coverage of the field of operation.

To address these issues, a multi-layered Monte Carlo simulation technique named incremental probabilistic simulation has been devised and implemented by the authors.

7.2.2   System and Component Rating

The structure of requirements on ISS/ADAS can be expressed in terms of the UML model given in Fig. 7.7. Requirements on the system (top level) are broken down into requirements on single components (lower levels) reflecting their interdependencies.

The authors are convinced that requirements on ISS/ADAS should be developed by following an iterative, cyclic top-down path through the use case diagram; a similar opinion is expressed in [5]. Since pure top-down development may easily lead to unsatisfiable requirements on the lower levels, the development should be accompanied by feasibility studies. Such feasibility studies may confirm assumptions and thus back the top-down approach to system requirements development.

The use case model lends itself to pointing out the distinction between system and component ratings. In terms of this model, a system rating (Fig. 7.7) measures the degree of fulfilment of system-level requirements (levels safety system and/or strategy). An example result of a system rating is "20% reduction of collisions".

In contrast to system ratings, a component rating (Fig. 7.7) reflects the degree of requirements fulfilment of the layer directly above. An example result of a sensor

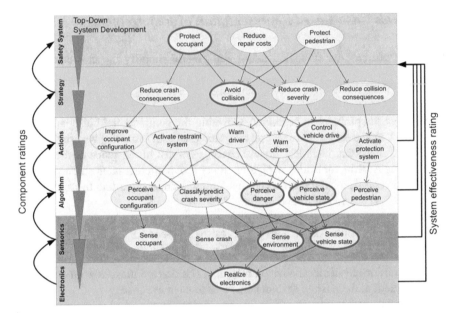

Fig. 7.7 UML use case model for integral safety with one function specification highlighted as an example

component rating might be "80% of objects are detected before the specified time to fire". While component ratings have some value, system ratings have far more significance for the ISS/ADAS system developer.

This line of thought motivates a very powerful technique, namely system rating with ideal component models; see Fig. 7.8. In early stages of development, due to the top-down paradigm, little is known about the properties of lower-level components. Assuming underlying components to be ideal (i.e. without physical or conceptual limitations, for example a sensor that always sees everything or an algorithm that always decides correctly) allows establishing theoretical performance limits. So, the implications of design decisions can be explored.

As an example, consider the development of a collision avoidance system. Assume this system should focus on front crashes. Following the top-down paradigm, one would begin the development by generating use cases that represent relevant accidents as well as non-accidents, e.g. using vehicle dynamics simulation software.

The next step is to define the system's behaviour in any use case. A first idea might be that the system should brake autonomously in order to avoid the collision. Assume a preliminary safety integrity level (SIL) rating suggests limiting the reduced velocity to, e.g. 30 km/h. When the collision cannot be avoided within this limit, the ego velocity should be reduced by 30 km/h at collision. Next, a precise definition of "front crash", including a quantitative evaluation criterion, is required in order to determine if a use case requires triggering the brake.

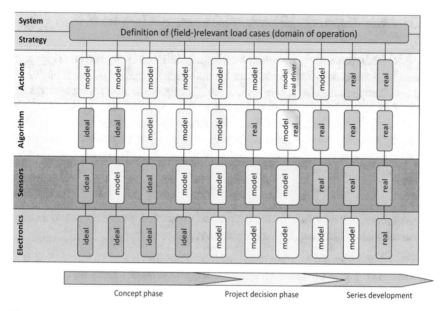

Fig. 7.8 Evolution of component models during system development

Having defined actions and domain of operation in this way, the theoretical limit of system effectiveness can be established.

Since limitations of algorithm, sensor and electronics have not yet been considered, a real system cannot be expected to actually meet this performance.

If the theoretical limit of system effectiveness is promising, the next step is to study the field of operation. This includes defining criteria that describe exactly which use case subset should be addressed by the ISS/ADAS. One might want to disable the autonomous brake when the driver brakes/accelerates or steers in order to avoid interfering with the driver's intent (compliant to the Vienna Convention on Road Traffic, see [3]). The effects of such system restrictions can be studied by including them in the system model and re-evaluating system effectiveness. Continuing the development, studies may include different sensor models or draft algorithms to further elaborate the system model. Along different project phases, the effectiveness ratings become more accurate as the system model converges towards the real system. Following this approach allows directing efforts to the most promising concepts. Since concepts and implementations of system components can be rated in a comparable way, it is also possible to identify components with the largest impact on system performance and to attribute performance losses to concept or implementation. Altogether, this approach massively front-loads development efforts and allows for a quantitative assessment of engineering decisions, both being highly valuable properties of a development methodology.

7.2.3 Data Mapping, Representativeness of Use Cases

It is evident that the test cases required for releasing ISS/ADAS depend on the specific system. An autonomous braking system has to be tested against different use cases than, for example, adaptive cruise control (ACC) or pedestrian warning systems. In early phases of top-down development, when a system specification is yet incomplete and has many degrees of freedom, a large variety of (potential) use cases is required and can be provided by simulation. When the system design matures, the use cases relevant for the specific system become more apparent. In early phases, development is primarily guided by simulation. As soon as the first physical system prototype becomes available, simulation may be supplemented by real tests.

While the cost per use case continually increases during development, the set of use cases that require explicit testing is expected to shrink due to better understanding of case relevancy. This raises the following issues:

(a) The system-specific relevant cases must be identified.
(b) Results of a small number of real tests have to validate results derived from a large number of simulations.
(c) A large number of simulations and a small number of real test cases have to be combined to form a line of argument that justifies a system release.

Resolving these issues requires intensive interdisciplinary collaboration. Issue (a) is already addressed in, for example, engine calibration by methods like "Online Design of Experiments" or "model-based calibration"; see [4] . Issue (b) can be tackled by approaches from areas such as statistical material modelling like "Operation Monitoring" [2]. Issue (c) requires the joint effort of consumer and standards organisations, product liability departments and engineers.

References

1. Aigner W, Kuhn A, Palau T, Marczyk J (2018) Expected systemic impacts on automated traffic from quantitative complexity rating. In: Proceedings of 25th ITS world congress, Copenhagen, Denmark
2. Bartolomé A, Palau T, Kuhn A, Rauh A, Mader H (2012) A machine learning based approach to large scale material model evaluation and iterative experimental design. In: European congress on computational methods in applied sciences and engineering, Vienna
3. Inland Transport Committee (1968) Convention on road traffic. United Nations Economic Commission for Europe
4. Hametner C, Jakubek S (2013) Local model network identification for online engine modelling. Inf Sci 220:210–225
5. Meitinger K-H (2008) Top-down-Entwicklung von aktiven Sicherheitssystemen für Kreuzungen (engl.: Top-down development of active safety systems for intersections). PhD thesis, Technische Universität München

Elvira Thonhofer has received her Masters Degree in Mechanical Engineering from theVien-naUniversity of Technology. Her research interests include traffic modelling, simulation and con-trol, intelligent transportation systems and automation. Elvira has worked on national research projects and her academic contributions are published in relevant journals and conferences.

José Carmona Ph.D. in Computer Science and Computer Science Engineer from the University of Málaga (Spain), is a development engineer at Andata GmbH. His research interest include com-putational intelligence, data mining and simulation in connection with traffic automation.

Chapter 8
Platoon Control Concepts

Alexander L. Gratzer, Alexander Schirrer, Sebastian Thormann, and Stefan Jakubek

Abstract Cooperative platoon control strategies utilise provided information from vehicle-to-everything (V2X) communication to reduce energy consumption and improve traffic flow and safety. In this chapter, a distributed control concept for cooperative platooning is developed that combines trajectory optimisation and local model-predictive control of each vehicle. The presented control architecture ensures collision safety by design, platoon efficiency and situational awareness with the option of exploiting V2X communication. The resulting platoon control performance is tested and validated in a realistic setting by utilising a co-simulation-based validation framework with detailed vehicle dynamics.

Keywords Distributed model predictive control (DMPC) · Platooning · Optimal control · String stability

8.1 Introduction

In order to enable high-performance, efficient and safe platooning control concepts, global properties such as surrounding traffic, infrastructure, platoon dynamics, road properties and route must be appropriately considered in the planning and optimisation of platoon trajectories. For the effective realisation of these movement patterns in vehicle control, the distributed or locally acting control on the individual vehicle level must be combined with the essential information from the broader, global context in a suitably prepared form. Cooperative platoon control strategies use provided information from V2X communication to reduce energy or fuel consumption, increase traffic flow and improve traffic safety. Thereby, local information and predictions can be shared with the entire platoon, thus improving the distributed control actions' effectiveness.

A. L. Gratzer (✉) · A. Schirrer · S. Thormann · S. Jakubek
Research Unit of Control and Process Automation, Institute of Mechanics & Mechatronics, TU Wien, Vienna, Austria
e-mail: alexander.gratzer@tuwien.ac.at

© The Author(s) 2022   105
A. Schirrer et al. (eds.), *Energy-Efficient and Semi-automated Truck Platooning*,
Lecture Notes in Intelligent Transportation and Infrastructure,
https://doi.org/10.1007/978-3-030-88682-0_8

8.2   Methodology Overview

A holistic control concept for cooperative platooning, illustrated in Fig. 8.1, was developed within Connecting Austria. Here, the *platoon coordinator* issues recommendations for actions to the platooning vehicles based on manoeuvre-specific trajectory planning, which is described in Sect. 8.4. The ego vehicle is locally controlled by model-predictive control (MPC). This MPC is specifically formulated so that a safe stop is always possible. Available information from the platoon coordinator as well as communicated predictions and agreements from the preceding and the following platooning vehicle are exploited for improved efficiency. The concept remains highly scalable due to its distributed control structure.

Each platooning vehicle implements a safety-extended local model-predictive controller wherein two optimisation problems are formulated: the *tracking problem* and the *fail-safe problem*. While the tracking problem aims at the tracking control of a reference trajectory, the fail-safe problem guarantees the possibility of a safe stop at any time. Due to this formulation, the platooning vehicles are always located in separated and thus safe position areas as illustrated in Fig. 8.1 top middle. The local *safeMPC* also implements two innovative methods for cooperative platooning and is explained in more detail in Sect. 8.5. First, a strategy is used to reduce the safely realisable distance between platooning vehicles by means of guaranteed temporarily limited brake actions; see Sect. 8.5.3. Second, an event-triggered communication scheme is used wherein the predicted ego trajectories are transmitted when necessary; see Sect. 8.5.4. Based on these methods, the platooning operation on slippery roads is considered and a special implementation variant (*explicit MPC*) is realised, which minimises the computational effort for real-time operation [10], see Sect. 8.5.5.

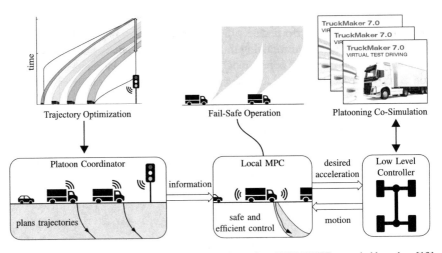

Fig. 8.1 Holistic control concept for cooperative platooning, ©2020 IEEE extended based on [13] with permission

Results for selected use cases are utilised for validation of the control concepts in Sect. 8.5. They are created by co-simulating detailed vehicle dynamics via the simulation software IPG Truckmaker® [6] whereby the control actions of the local MPCs are computed via MATLAB® [11]. A more detailed description of the co-simulation architecture is given in Sect. 8.3.

8.3 Co-simulation-Based Validation

For validation of the developed control concept, a co-simulation of the control algorithms with detailed vehicle simulations has been designed: with multiple instances of the simulation software IPG Truckmaker®, the vehicle dynamics are simulated in detail, and thus, a realistic test environment is achieved. Truckmaker® simulates multi-body dynamics enhanced with gear box, clutch, engine, and tyre models for each individual truck. The control law computation, communication, and synchronisation are realised via a central MATLAB® session in this co-simulation set-up. The complex nonlinear truck system dynamics considerably deviate from the idealised control design models, especially the drive train dynamics as well as the dynamic response to desired acceleration inputs. The implementation uses a central MATLAB® session which coordinates the individual Truckmaker® vehicle simulation instances and guarantees real-time capability for several trucks; see Fig. 8.2. Each vehicle instance is connected to a MATLAB®/Simulink®-worker, who also communicates with the platoon coordinator. The MPC problems are formulated and solved by parallel computing using the YALMIP toolbox [8] and Gurobi® [1] as optimisation problem solver. The simulation can also be connected to a force-feedback steering wheel, which allows to experience the platoon behaviour in traffic in a human-in-the-loop fashion. This way, new functionality of the platoon controller can be tested and evaluated effectively.

The functionalities of the platoon control concepts described hereafter are implemented and tested in the described co-simulation framework and driving simulator solution.

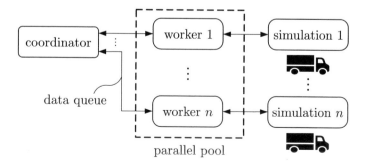

Fig. 8.2 Co-simulation architecture used for validating selected use cases and features

8.3.1 String Stability Considerations

String stability in a platoon is a property of platoon dynamics, expressing whether disturbances injected at the front of the platoon grow or decrease as they are propagated towards the platoon tail. It is especially important if platoons with short inter-vehicle distances and/or many vehicles should be built. Only string-stable platoon dynamics allows well-performing long platoons to be formed. String stability can be characterised by various approaches as surveyed in [4]. In particular, one group of string stability criteria is based on the $\mathcal{H}_\infty$ system norm for the transfer from the predecessor's velocity v_{pre} to the resulting ego vehicle's velocity v_{ego}, formally given by

$$\|G_v(z)\|_\infty \leq 1 \Leftrightarrow \text{predecessor-follower-string stable},\tag{8.1}$$

in which $G_v(z)$ denotes the string stability transfer function from v_{pre} to v_{ego} in the case of linear discrete-time system dynamics.

For complex, nonlinear system dynamics, the $\mathcal{H}_\infty$-string stability criterion (where (8.1) represents the linear-dynamic case) is generally defined as

$$\sup_{\|v_{pre}(t)\|_2 \neq 0} \frac{\|v_{ego}(t)\|_2}{\|v_{pre}(t)\|_2} \leq 1 \Leftrightarrow \text{predecessor-follower-string stable}.\tag{8.2}$$

Therein, $\|\cdot(t)\|_2$ denotes the signal energy norm of the corresponding time signal. Unfortunately, this criterion cannot directly be determined in practice because all possible trajectories would have to be evaluated.

However, focusing our attention on a particular manoeuvre with finite-energy error velocity signals (both v_{pre} and v_{ego} approach the same reference velocity v_{ref} as $t \to \infty$), the manoeuvre's signals can be evaluated and a lower bound for the $\mathcal{H}_\infty$ gain can be obtained:

$$\frac{\|v_{ego}(t) - v_{ref}\|_2}{\|v_{pre}(t) - v_{ref}\|_2} > 1 \Rightarrow \textbf{not} \text{ predecessor-follower-string stable}.\tag{8.3}$$

Analogously, energy ratios in accelerations or distance errors can be formulated and utilised to interpret simulation results for string stability properties.

In [7], a set of tools for the assessment of string stability properties of platooning controllers is presented. Both linear and nonlinear (simulation-based) system dynamics can be used as an evaluation basis for the influence on congestion formation in real traffic. Both analytic and empirical settings are proposed therein.

Figure 8.3 illustrates string-stable (*left*) versus not string-stable (*right*) platoon dynamics in a transient manoeuvre. The platoon initially drives at $v = 80\,\text{km/h}^{-1}$ with an inter-vehicle distance of $d_\infty = 18\,\text{m}$ respective $d_\infty = 15\,\text{m}$, and an external non-platoon vehicle induces a brake pulse. In the string-stable case, the resulting disturbances decay as they are propagated towards the platoon tail vehicle, whereas in the other configuration, the disturbance amplitudes significantly increase along the

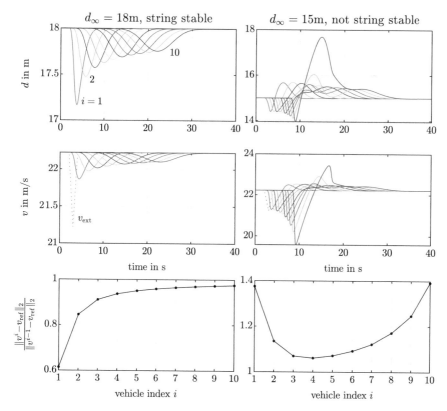

Fig. 8.3 Inter-vehicle distance d, velocity v and evaluated Eq. (8.3) for initial inter-vehicle distance $d_\infty = 18$m (*left*, string stable) and $d_\infty = 15$m (*right*, not string stable)

platoon. Note that each vehicle nevertheless shows a stable response. Figure 8.3 also shows the empirical string stability measure (8.3) for the cases when string stability is present (*left*) or not (*right*), respectively.

Further analysis regarding string stability of cooperative vehicle platoons with consideration of collision safety constraints, as implemented in the proposed *safeMPC* concept, can be found in [5].

8.4 Trajectory Optimisation Methodology

Trajectory planning enhances the performance of the implemented distributed control scheme from a global perspective by incorporating information obtained from the infrastructure, for example speed limits, road conditions or green light timing information.

In order to enable group manoeuvres, such as a simultaneous start-up in front of a traffic light or an efficient transition to manual driving when approaching a hazardous location, a trajectory optimisation task is carried out by a *platoon coordinator* instance. This platoon coordinator is understood to be an entity that issues recommendations for actions to the platooning vehicles based on additional information obtained through V2X communication including positions and velocities of the controlled platooning members. These recommendations include position trajectories as well as desired velocities and desired minimal inter-vehicle distances. The optimised trajectories, which are planned in a way that safety and optimality objectives are achieved, are used as reference values for the distributed local MPCs, which aim to track these planned trajectories and also take into account the current traffic situation in order to calculate collision-safe control inputs.

In practice, the implementation of the platoon coordinator can be done, for example, at the platoon leader (first vehicle of the platoon), but also an implementation in a road-side unit (RSU) or a completely distributed implementation is possible.

8.4.1 Optimisation Problem Formulation

The trajectories are planned in such a way that application-specific safety and optimality goals are achieved. The underlying optimisation problem is generally of the form

$$U^* = \arg\min_U J(U; \theta)$$

such that the restrictions

$$g(U; \theta) \leq 0$$

are satisfied. Thus, the scalar cost function $J(U; \theta)$ is to be minimised by an optimal choice of the decision variables U at fixed values of the parameters θ while at the same time the constraints modelled as a vector-valued constraint function $g(U; \theta)$ must be obeyed. Thus, the optimal values of the decision variables U^* are exactly those which minimise the costs and simultaneously satisfy the given constraints. The trajectory optimisation sequentially solves quadratic programming problems with linear constraints. For the concrete applications in the project, the general form of the optimisation problem will be interpreted illustratively in the following. The decision variables U are the control inputs of the platooning vehicles, which are understood as acceleration values at a range of time steps covered by the optimisation horizon. The cost function $J(U; \theta)$ is composed of application-specific terms, such as costs due to the control input values, the time losses during passing an intersection or the deviations from desired inter-vehicle distances. The prioritisation of these terms is achieved by weighting factors in the objective formulation of the cost function. The restrictions $g(U; \theta)$ take into account, among other things, simplified vehicle dynamics with limited acceleration (characteristic curve dependent on

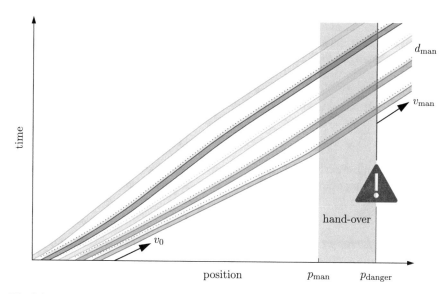

Fig. 8.4 Optimisation of a handover to the human driver in front of a hazardous location (construction site) at p_{danger} with demanded velocities v_{man} and distances d_{man} during the handover phase

speed) and minimum required distances in terms of space and time gaps. In order to effectively use the trajectory optimisation as a tool set, a formal interface definition was formulated and implemented, which simplifies the definition of the parameters θ. Also, semantic tests with respect to the given problem specification and the calculated results are carried out to allow safely automated parameterisation and Monte Carlo-type simulation studies.

Since the results of the trajectory planning task have only informative character, the optimisation problem presents a less time critical operation [12].

The two scenarios *use case 2: truck platoon approaching a hazardous location*, as illustrated in Fig. 1.2, and *use case 4: truck platoon crossing an intersection*, as depicted in Fig. 1.4, are considered in particular in the following Sects. 8.4.2 and 8.4.3, respectively.

8.4.2 Trajectory Optimisation for Approaching a Hazardous Location

Figure 8.4 shows an example for the case of a platoon disbanding and the subsequent handover to the human driver when approaching a *hazardous location*, such as a construction site with a reduced speed limit. In front of the construction site, at a distance of 600 m from the initial platoon location, the platoon should expand to an inter-vehicle distance of $d_{man} = 50$ m, and a speed of $v_{man} = 60$ km/h^{-1} should be maintained during the handover phase. The vehicle lengths are 15 m. In order to allow a sufficient time span for the handover to the human driver, this required platoon state has to be realised already 5 s before reaching the construction site. Since the velocity during the handover phase is assumed to be kept constant at v_{man}, the handover for each vehicle begins at the position

$$p_{man} = p_{danger} - v_{man} \, T_{man} \, .$$

In this optimisation problem, an energy-efficient trajectory is sought which meets the required platoon state with high accuracy during the handover phase.

8.4.3 Trajectory Optimisation for Crossing an Intersection

Given the use case of a platoon crossing a traffic light-controlled intersection, it is assumed that the platoon coordinator receives accurate real-time information of the next green light phase via infrastructure-to-vehicle (I2V) communication. By carrying out a centralised but simplified trajectory optimisation where time and energy consumption are minimised, the platoon coordinator is able to plan and recommend efficient position trajectories as well as minimum inter-vehicle distances and velocities for crossing the intersection. This functionality represents an extension to Green Light Optimal Speed Advisory (GLOSA) features already available in autonomous driving use cases [3].

The trajectory optimisation problem set-up is illustrated in Fig. 8.5. Here, a so-called *optimal drive* is sought, which allows the platoon to pass the intersection with efficient control inputs as early as possible during the green light phase of the traffic light. In detail, two auxiliary problems are solved: the *maximum drive* of the platoon leader is the fastest possible manoeuvre at any time until the beginning of the green light phase, so that the traffic light is not run over when still red. The *minimum drive* of the platoon tail (last vehicle of the platoon) is the slowest possible manoeuvre at any time until the end of the green light phase, so that the traffic light is reached in the last moment of the green phase. The optimal drive is then solved using the results of the minimum and maximum drives as bounding constraints [12].

However, the boundary conditions of such traffic optimisations are only partly known. For example, if non-platooning vehicles are present and their future behaviour is uncertain, then continuous adaptation of these planned trajectories is necessary. Utilising event-triggered vehicle-to-vehicle (V2V) communication, the platoon is able to efficiently adapt its controlled motion to the sudden presence of a non-platooning vehicle and thus to realise situational awareness as will be described in Sect. 8.5.4.

Fig. 8.5 Trajectory optimisation for the use case of a truck platoon crossing an intersection. The individual coloured areas represent the occupied space of each vehicle respective the demanded velocity-dependent time gaps

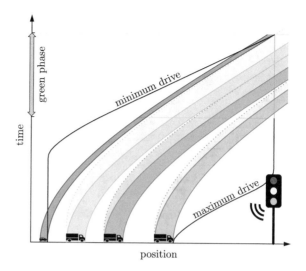

position

8.5 Distributed Model-Predictive Platoon Control

The basic concept of MPC is to use a dynamic model to predict the behaviour of the controlled system and recurrently optimise the control input in order to obtain an optimal predicted system response, as introduced in the textbook [9] and illustrated here in Fig. 8.6. The dynamic model used to calculate predictions, the *prediction model*, can be based on known physical laws, empirical data and/or expert knowledge. Usually, the predictions are considered only over a limited future time interval (up to the *prediction horizon*) to render the involved calculations feasible in real-time. The meaning of an optimal system behaviour has to be specified. For this purpose, the predicted system behaviour (predicted state and predicted input) is assessed from the current point in time to the prediction horizon. This assessment is implemented by means of a cost function which quantifies the attainment of specified control goals, such as accurate reference tracking, energy-efficient actuation or time optimality. Constraints of the controlled system, such as permitted inter-vehicle distances, or speed limits, can be directly incorporated. Eventually, an optimisation problem of the form (8.4.1)–(8.4.1) is solved recurrently to calculate the currently applied input. Thereby the value of the cost function is minimised while the given constraints are obeyed.

In *distributed MPC*, control actions are only based on a subset of the overall system information. Reasons for this may include unavailable information of distant subsystems, reduced communication effort and/or reduced computational complexity. Platoon control is a typical application domain for distributed MPC. For individual vehicles, local measurements are available. Additionally, in cooperative platoon control, neighbourhood information may be communicated. The distributed MPC problem is now characterised by control goals and/or constraints which involve non-

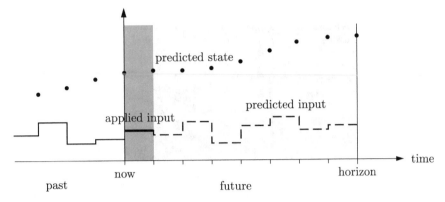

Fig. 8.6 Basic concept of model-predictive control

local system states, while the control actions are based on a local information scope and communicated information.

8.5.1 Safe-by-Design Local MPC Formulation

Each platoon vehicle needs a local controller to track its preceding vehicle safely and efficiently. The local platooning vehicle implements a model-predictive controller (MPC), which is formulated considering collision safety explicitly. These concepts have been developed in [13] and [10] and are summarised in the following.

The local MPC structure follows a safe-by-design paradigm and at the same time takes communicated information from the platoon coordinator as well as from the front vehicle into account to improve efficiency. Figure 8.7 illustrates the safety aspects of the MPC formulation. Two optimisation problems are formulated (Fig. 8.7, *left*), the *tracking problem* and the *fail-safe problem*, which are coupled over a time span ("tolerance time") T_{tol}. While the tracking problem aims to track a particular reference trajectory (determined by the desired velocity, predecessor predictions and corresponding gap policies), the fail-safe problem ensures the feasibility of a safe (i.e. collision-free) stop behind its predecessor at all times. Consequently, the platooning vehicles are always kept in separated and thus safe areas (Fig. 8.7, *right*).

In [13], two additional, innovative methods for cooperative platooning have been developed: first, a strategy was developed to reduce the safely realisable distance between platooning vehicles by committing to temporarily limited brake authority. Second, an event-triggered communication scheme for the efficient transmission of local MPC predictions was developed, which allows for an early response when a control intervention is necessary. Based on this fundamental control structure, in [10] the platooning operation on slippery roads was considered and a special implemen-

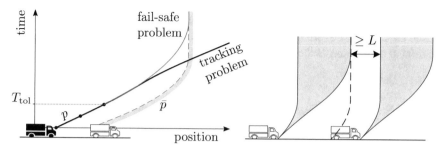

Fig. 8.7 Safety aspects of the MPC formulation, ©2020 IEEE adapted from [13] with permission: Keeping a fail-safe trajectory feasible at all times, the coupled tracking problem can be designed separately, and collision safety is guaranteed (*left*). Each vehicle can realise a safe set of trajectories which is disjoint from the other vehicles' trajectories (*right*).

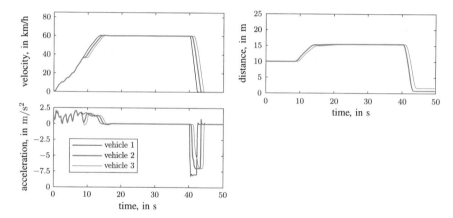

Fig. 8.8 Simulation scenario: emergency braking

tation variant (explicit MPC [2]) was realised, which minimises the computing effort for real-time operation.

8.5.2 Validation of Collision Safety via Co-simulation

The presented control concept meets realistic requirements regarding the robustness against model errors, which was validated by co-simulations using the commercial software IPG Truckmaker® for the simulation of detailed vehicle dynamics. Two simulation scenarios will now be explained by way of example. Figure 8.8 shows the simulation results of the controlled platoon during emergency braking of the platoon leader at time $t = 20$ s. All platoon vehicles come to a safe (i.e. collision-free) standstill.

In addition to these co-simulation studies, a human-in-the-loop driving simulator based on the same co-simulation framework has been developed. It allows to gather first-hand experience of the platoon control features from the driver's perspective.

8.5.3 Safe Reduction of Inter-vehicle Distances

The safety constraint illustrated above is significantly affected by the bounded or estimated braking authority of the predecessor and the tolerance time. To safely reduce inter-vehicle distances, the preceding vehicle could commit to a reduced deceleration magnitude. This *hold-back* strategy for temporarily reduced braking authority is detailed in [13]. Using absolute time stamps of the expiration time of this commitment, the approach becomes robust against a loss of V2V communication.

Use case 2 has already been discussed in trajectory planning. In terms of use case 2, moreover, the hold-back feature is relevant: on a free highway, the predecessor (ultimately the leader vehicle of the platoon) can commit to significantly reduce deceleration magnitudes for relevant time spans safely, which, in turn, allows the platoon to realise tight inter-vehicle gaps safely. When the surrounding traffic situation changes, as for example in the approach to a hazardous location (Use Case 2), the hold-back phase safely expires, and the platoon automatically and safely opens up to larger, safe inter-vehicle gaps.

Figure 8.9 shows the platoon behaviour in case of this limited brake authority (hold-back strategy). With active hold-back (time ranges marked in grey), each platooning vehicle guarantees a defined, reduced deceleration magnitude bound until the expiration of a communicated time, which is regularly updated and thus postponed into the future. This allows the safe minimum distance between the platooning vehicles to be significantly reduced. If the hold-back strategy is now deactivated, for example by interrupting the communication (comm. lost in Fig. 8.9), the platoon will automatically trigger an always safe expansion of the platoon.

8.5.4 Situation-Aware Platoon Behaviour via V2V-Communication

This section outlines a corridor-based platoon communication strategy to realise situation awareness with parsimonious V2V communication [13]. This feature has direct application in use case 4 and provides a flexible, generic and efficient solution to the case when the platoon has to react appropriately to individual, non-platoon vehicles on the road.

We illustrate the proposed method at the case when an *individual, non-platooning vehicle is present in front of a traffic light* as shown in Fig. 8.10. Initially, the platooning vehicles track the planned trajectories (dashed lines), compare Sect. 8.4.3.

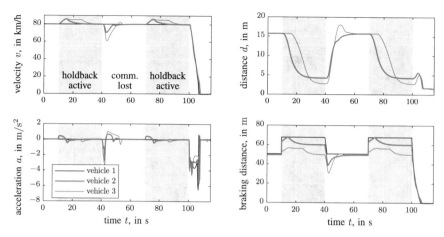

Fig. 8.9 Use case 2: truck platoon approaching a hazardous location (hold-back, communication loss at $t = 40$s, emergency brake at $t = 100$ s)

Since a non-platooning vehicle (dash-dotted line), waiting at standstill before the intersection, was not considered in the trajectory optimisation, the platoon adapts to this unpredicted traffic situation: the individual vehicle's motion is predicted based on current sensor data measured by the leader and considered in solving its predictive control problem. The leader communicates its own predicted trajectories to its followers if this data is sufficiently different to the last transmitted trajectories. As a result of these automated event-triggered prediction updates within the platoon, an efficient braking manoeuvre as well as a simultaneous start-up is realised automatically in this distributed control setting without centralised guidance. This case demonstrates robust situation awareness capabilities of the outlined control concept. The necessary communication bandwidth is kept low. Also, the communication of accurate predictive information within the platoon establishes string-stable behaviour with respect to the updated manoeuvre, as seen in the essentially congruent trajectories around $t = 15$ s.

8.5.5   Consideration of Varying Road Conditions

The ideas of the safety-extended distributed model-predictive platoon control concept shown in [13] have been utilised to design an explicit distributed platoon MPC for varying road slip conditions in [10]. The shown controller provides collision safety and realises platooning functions under the assumption that the current friction coefficient of the road is available or estimated. The explicit control law is a precomputed form of the MPC which can be constructed if the number of problem parameters (initial state, references, constraint parameters) is sufficiently low. Then, a lookup surface in this parameter space is constructed which represents the solution

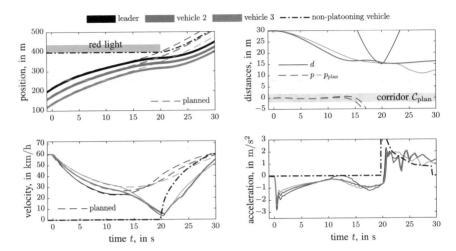

Fig. 8.10 Simulation results of the cooperative platooning strategy for the case of an present non-platooning vehicle in front of a traffic light. It can be seen that the platoon adapts to the unpredicted new situation in an optimal way by diverging from the planned trajectories for about 20 s

of the underlying optimisation problem directly. It was shown that the main features of the safe-by-design local MPC outlined in [13] can be simplified in a suitable way to obtain a tractable explicit MPC formulation (see [10]). To achieve this simplification, a non-uniform sampling of the optimisation horizon of the discrete-time MPC formulation has been considered, and the tracking of vehicle reference velocities and inter-vehicle distances has been parameterised in an efficient way through a corresponding preprocessing step. Also, the collision safety constraints have been formulated in a simplified and approximated fashion. The resulting control performance and collision safety have been tested in platoon co-simulation studies with detailed vehicle dynamics. Finally, the platoon behaviour on a dry road as well as on a slippery road has been validated in co-simulations.

8.6 Conclusion

A holistic predictive control concept has been developed that involves manoeuvre planning and optimisation, safe-by-design distributed model-predictive platoon control of the vehicles, and a detailed co-simulation validation framework.

Trajectory planning methods are utilised to generate efficient real-time trajectories of each platooning vehicle (position, velocity and acceleration over a defined time interval from the current time into the future). A general formulated based on optimal control has been formulated which allows to solve many relevant platoon manoeuvres and scenarios, such as the approach to a hazardous location, or the task of crossing an intersection efficiently and safely. Real-time information from the road infrastructure is incorporated, and relevant constraints such as safety distances and traffic rules are considered. The resulting trajectories serve as references for the distributed model-predictive platoon vehicle controllers which control the individual vehicles accordingly.

The proposed vehicle control architecture safely and efficiently controls the platoon vehicles, incorporates infrastructure information in real time into the control task and provides collision safety guarantees. Moreover, suitable measures are outlined which allow to exploit V2V communication between platooning vehicles to safely reduce inter-vehicle distances and improve platoon efficiency.

The validation framework is based on the co-simulation of all platoon vehicles, each simulated with detailed vehicle dynamics by a physically realistic, industry-grade vehicle simulator and equipped with the novel distributed MPC and communication functions. The range of platoon control functionalities is further extended by parameterising the controller with a real-time road friction estimate, so that safety is still ensured if road conditions deteriorate.

The outlined collection of methods, tools and control functionality provides a versatile and comprehensive basis for intelligent, situation-aware and safe platooning, directly applicable in realistic platooning scenarios.

References

1. Manual Gurobi Optimizer Reference (2018) Version 8:1
2. Bemporad A, Morari M, Dua V, Pistikopoulos EN (2002) The explicit linear quadratic regulator for constrained systems. Automatica 38(1):3–20
3. CAR 2 CAR Communication Consortium (2019) Guidance for day 2 and beyond roadmap
4. Feng S, Zhang Y, Eben Li S, Cao Z, Liu HX, Li L (2019) Definitions and analysis methods. String stability for vehicular platoon control. Annu Rev Control 47:81–97
5. Gratzer A, Thormann S, Schirrer A, Jakubek S (2021) String stability of cooperative vehicle platoons with consideration of collision safety constraints. submitted, under review
6. IPG Automotive GmbH. CarMaker release notes 7.0.3
7. Kalteis C, Thormann S, Schirrer A, Jakubek S (2020) Efficient methods to assess linear and nonlinear automotive platoon control stability and performance. In: XI international conference on structural dynamics:748–757
8. Löfberg J (2004) YALMIP: a toolbox for modeling and optimization in MATLAB. In: Proceedings of the IEEE international symposium on computer-aided control system design, 284–289
9. Rawlings JB, Mayne DQ, Diehl M (2017) Model predictive control: theory, computation, and design. Nob Hill Publishing Madison, WI, 2 edition
10. Schirrer A, Hanis T, Klauco M, Thormann S, Hromcik M, Jakubek S (2020) Safety-extended explicit MPC for autonomous truck platooning on varying road conditions. IFAC world congress
11. The MathWorks Inc. MATLAB R2018b, 2018
12. Thormann S (2020) Cooperative platooning—development and co-simulation-based validation of distributed model predictive control methods for safe and efficient cooperative platooning. Master's thesis, TU Wien
13. Thormann S, Schirrer A, Jakubek S (2020) Safe and efficient cooperative platooning. IEEE Trans Intell Transp Sys

Alexander L. Gratzer has been a Project Assistant with the Institute of Mechanics and Mechatronics, TU Wien, since 2019 and currently works toward the Ph.D. degree. Alexander studied mechanical engineering and is involved in international research projects and teaching of graduate-level lectures. His research interests include simulation, optimisation, and control of complex industrial systems.

Alexander Schirrer has been a Postdoctoral Researcher and Teacher of graduatelevel lectures with the Institute for Mechanics and Mechatronics, TU Wien, since 2011. His research interests include modelling, simulation, optimisation, and control of complex and distributed-parameter systems.

Sebastian Thormann has been a Project Assistant with the Automation and Control Institute, TU Wien, since 2021 and currently works toward the Ph.D. degree. He studied mechanical engineering and was a student member of the project team with the Institute for Mechanics and Mechatronics, TU Wien, from 2017 to 2020. His research interests include modelling, simulation, optimisation, and control of complex and distributed-parameter systems.

Stefan Jakubek is Professor and Head of the Institute of Mechanics and Mechatronics, TU Wien. From 2007 to 2009, he was the Head of Development for Hybrid Powertrain Calibration and Battery Testing Technology with AVL List GmbH, Graz, Austria. His research interests include fault diagnosis and system identification.

Chapter 9
Scenario-Based Simulation Studies on Platooning Effects in Traffic

Andreas Kuhn, José Carmona, Elvira Thonhofer, and David Hildenbrandt

Abstract This chapter outlines the portfolio of simulation campaigns that have been carried out to thoroughly study the effects of platooning in the traffic system. The approach outlined in Chap. 7 is utilised to quantify typical platoon trajectories and manoeuvres in highway settings as well as in urban intersection scenarios. The addressed studies do not yield a single result, but instead depend on many parameters (such as platoon spacing/gap policy, surrounding traffic density and speed and many more) and are investigated in terms of the results' sensitivities on these parameters. This approach allows one to draw meaningful conclusions despite the inherent uncertainty and spread of the influencing parameters. By using representative conditions, the resulting KPI distributions are evaluated and interpreted. Considering real traffic parameters, such as density, truck share, distances, speed and their empirical distributions and restrictions on the assumed "degree of connectivity" of trucks, maximum platoon length, an estimation of the real achievable traffic efficiency and the potential for improvement relative to the current status can be calculated.

Keywords Platooning · Traffic micro-simulation · Analytical traffic models · Effectiveness assessment · Scenario management

9.1 Intersection Scenarios

In this section, selected intersection scenarios are discussed to exemplify the application of some of the methods described in Chap. 7.

A. Kuhn · J. Carmona (✉) · E. Thonhofer · D. Hildenbrandt
Andata Artificial Intelligence Labs, Hallein, Austria
e-mail: jose.carmona@andata.at

A. Schirrer et al. (eds.), *Energy-Efficient and Semi-automated Truck Platooning*,
Lecture Notes in Intelligent Transportation and Infrastructure,
https://doi.org/10.1007/978-3-030-88682-0_9

9.1.1 Green Time Extension

Green time extension is a strategy for the platoon to cross the intersection without being split by the traffic light phasing. Conceptually, the intersection's infrastructure receives information about the incoming platoon and extends the corresponding green phase within predefined limits. This reduces the lost time of the platoon, because the red phase is potentially avoided, as well as the platoon splitting. There is an additional advantage, namely the overall energy saving, because stopping a truck is more energetically expensive than stopping a car. Nevertheless, there is a negative effect for the traffic flow for some of the other streets in the intersection because their corresponding green time is reduced. Micro-simulation models and scenario variations allow to quantify this conflict numerically.

Micro-simulation Model

In order to quantify all these effects, it is necessary to develop a detailed micro-simulation model, as introduced in Sect. 7.1.3, containing the intersection geometry, traffic signs and traffic lights.

In order for the results to be representative, it is also necessary to quantify the initial and boundary conditions in the form of inflows, traffic composition or routing decisions. This can be achieved using the methods described in Sect. 7.1.1.

Systematic Scenario Simulations

Once a validated model is available, it can be used for systematic scenario simulations as described in Sect. 7.1.4. A high number of variants are generated by varying initial conditions, signal plans and control-specific parameters, such as the maximal green time extension. The specific intersection to be modelled is also subject to variation. These variants are then simulated, and the resulting key performance indicators are evaluated (as described in Sect. 7.1.5). As part of the evaluation, the benefits for platoons in terms of reduced lost time as well as the negative effects on other streets can then be analysed.

9.1.2 Coordinated Drive-Away

A further intersection scenario is the coordinated drive-away when the traffic light is initially red and turns to green during the scenario. The coordinated drive-away has been assessed as a strategy to minimise the accumulated reaction time. Trajectory optimisation methods can be used to calculate the optimal trajectories of the platooning trucks that comply with the given restrictions, such as minimal distance and acceleration limits (see Chap. 8). This assumes that the signal plan is known to the platoon, that is, it is known how many seconds are left until the green phase starts. In the resulting scenario catalogue, several parameters are varied, such as acceleration limits, initial spacing between the vehicles and others.

In this scenario, the focus lies on the evaluation of the minimised lost time in comparison with the lost times that result from the human driving behaviour, specifically the reaction times. It is thus necessary to quantify reaction times at intersections by using naturalistic driving studies, as outlined in Sect. 7.1.2.

9.1.3 Optimisation of Speeds and Distances Inside the Platoon

Another intersection scenario consists of the optimisation of the speeds and distances inside the platoon by making use of phase plan information. That is, the platoon can drive faster or with less distance between its vehicles in order to reach the green phase, assuming a fixed signal plan.

This strategy reduces the lost time of the platoon, since the red phase can be avoided. The traffic flow is also improved, making a better use of the fixed signal plan. Since platoon stopping or splitting is avoided, the strategy also generates advantages from an energetic point of view.

Assumptions

This scenario assumes that accurate information on the signal plan is known to the platoon, in order to be able to optimise the trajectories accordingly. This information could be sent to the platoon by means of I2V communication.

Systematic Scenario Simulations

In the scenario catalogue, the initial platoon positions and distances are systematically varied, as well as signal plans, vehicle lengths, vehicle characteristics such as acceleration ranges, and control parameters such as speed limits or minimal distances between the vehicles. Another aspect which is subject to variation is the minimal distance to the traffic light at which the trajectories optimisation can begin. It is important to vary this factor, since this is equivalent in evaluating the strategy with different communication ranges, and it allows to quantify the efficacy of different communication ranges in the context of this strategy.

9.2 Application of Analytic Approaches: Highway Throughput Based on Platooning Headway

This section proposes an example for a scenario in which analytic approaches can be taken. The influence of reduced platooning headway on traffic efficiency is assessed in two stages. First, a theoretical upper limit for the compression of the traffic under simplified assumptions is determined. Second, relevant parameters are sampled from broad ranges, and stochastic results are generated. These can be evaluated either in a very general way or specifically and a posteriori for a set of parameters whose values

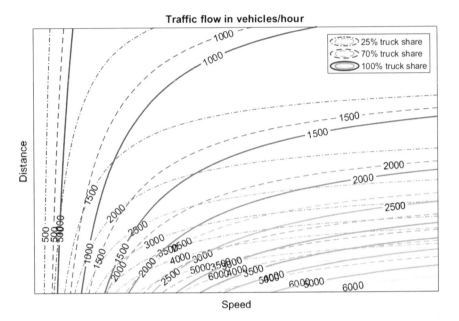

Fig. 9.1 Traffic flow as a result of specific combinations of driving speed and intra-platoon distance for three levels of truck share

are unknown at the time of writing but will gradually become available as technology advances.

9.2.1 Analytical Models for the Traffic Throughput

The theoretical upper limit can be calculated based on mixed platooning (i.e. between all vehicle types) and full penetration with platooning capability. There is a functional relation between speed v, platooning distance h_p and traffic throughput ϕ_0,

$$\phi_0 = \frac{v_0}{(L_{veh} + h_p)} \tag{9.1}$$

where L_{veh} is the mean length of the vehicles. By using such analytical relation, it is possible to assess the theoretical upper bound on traffic efficiency that we can achieve with a penetration rate of 100% and mixed platooning. Both assumptions are not realistic in the foreseeable future, so the results explicitly serve as a theoretical upper limit. Figure 9.1 shows the theoretical traffic flow that can be achieved for three levels of truck share.

Using real-world data allows to use realistic truck shares in the traffic, as well as vehicle length distributions by vehicle type. It is then possible to calculate yet more realistic upper limits if arbitrary platooning vehicles penetrations are allowed. The following relation can be formulated:

$$\rho = \frac{1000}{\alpha\beta(L_{truck} + h_p) + \alpha(1 - \beta)(L_{truck} + h) + (1 - \alpha)(L_{car} + h)} \qquad (9.2)$$

where

- ρ is the density (vehicles/km).
- L_{truck} is the length of the trucks.
- L_{car} is the length of the cars.
- α is the proportion of trucks.
- β is the proportion of platooning trucks over the total number of trucks.
- h_p is the distance between the platooning vehicles.
- h is the distance between the non-platooning vehicles.

The resulting traffic flow is then defined as $\phi = \rho v$.

9.2.2 Stochastic Variations

By using an analytical model as above, it is possible to generate a large number of scene variants that include systematic variations of platooning share, traffic composition and vehicle lengths based on real-world data, as well as normal and platooning distances. It is possible to generate scenarios where all of the vehicles or only the trucks can be part of a platoon.

9.3 Theoretical Lower Limits on Intra-platoon Distance

A further example of the application of the methods of Chap. 7 is the study of the minimal intra-platoon distances for collision avoidance. In this scenario, non-platoon traffic participants are not considered. A stable moving platoon is assumed (constant, suitable speed, stable driving condition of all vehicles). More specifically, in this scenario there is a leading truck and a follower truck.

The leading truck brakes, and the reason of braking (obstacle, other vehicle, etc.) is not of interest here. This is communicated to the follower vehicle in the form of V2V messages. After a given communication delay, the follower truck starts to brake in order to try to avoid the collision. Safe braking distance considerations are illustrated in Fig. 9.2.

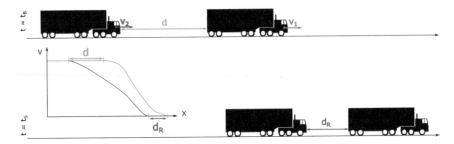

Fig. 9.2 Safe braking within a platoon. Starting from an initial distance d and initial speeds v_1 and v_2 at time t_B, each truck decelerates according to its deceleration capacity. A positive residual distance d_R at time t_S indicates that a collision can be avoided

9.3.1 Scenario Definition

Several use cases can be defined, depending on the braking profile of the leading truck:

- Normal braking.
- Full braking.
- Emergency braking.

Following the methodology proposed in Chap. 7, the basic scenarios for the analysis have to be defined. A basic scenario for any of these subcases comprises a combination of:

- Street geometry, curvature profile.
- Initial speed.
- Initial positions.
- Vehicle characteristics, including braking limits and mass.
- Friction between street and tyre.

9.3.2 Evaluation of KPIs

The evaluation parameters that we use to evaluate the results (see Sect. 7.1.5) are the collision probability. In order to be able to quantify these probabilities, several simulations of the scenes are carried out varying the time delay with which the follower truck begins the braking manoeuvre (relative to the leading truck), as well as the braking profile of the follower truck (including acceleration limits and the braking duration). In each of the simulations, a crash can occur. The collision probability is thus the ratio between collisions observed and number simulations for each base scene.

Andreas Kuhn studied Technical Mathematics and Mechanical Engineering at the Vienna University of Technology. There he also awarded his Ph.D. for the simulation of special satellite dynamics. He now works for more than two decades in several positions and roles in the fields of automotive safety, automated driving and traffic automation with an steady focus on virtual development procedures and the safe application of softcomputing methods.

José Carmona Ph.D. in Computer Science and Computer Science Engineer from the University of Málaga (Spain), is a development engineer at Andata GmbH. His research interest include computational intelligence, data mining and simulation in connection with traffic automation.

Elvira Thonhofer has received her Masters Degree in Mechanical Engineering from the Vienna University of Technology. Her research interests include traffic modelling, simulation and control, intelligent transportation systems and automation. Elvira has worked on national research projects and her academic contributions are published in relevant journals and conferences.

David Hildenbrandt has received a Master's Degree in Physics from the University of Vienna. He works as a development engineer and data scientist at ANDATA GmbH and his research interests include computational intelligence, stochastic and nonlinear dynamics as well as modelling and simulation of intelligent transportation systems.

Chapter 10
Energy-Efficient Internet of Things Solution for Traffic Monitoring

Thomas Hoch and Theodorich Kopetzky

Abstract Recent progress in video-based vehicle sensors allows for a detailed observation of road users on intersections in urban areas. By combining the measured real-life traffic situation with thorough traffic simulations, a cooperative system design for the dynamic management of traffic flow including vehicle platoons is possible. In this chapter, we describe our video-based traffic flow estimation system that we installed at a three-way intersection in the small city of Hallein, Austria. We show that the installed system is able to collect comprehensive information about the traffic situation in near real time, and that this information can be used to estimate traffic density and flows of cars and trucks with high precision.

Keywords Edge AI · IoT · Traffic flow analysis · Object detection

10.1 Introduction

The special focus of the Connecting Austria project on the road infrastructure and innovative C-ITS services for level 1 truck platooning as outlined in [20] made the development of a video-based traffic flow estimation system for the assessment of traffic efficiency and safety of platooning in urban areas necessary. The scenario-based evaluation of *use case 4: truck platoon crossing an intersection* is carried out comprehensively for realistic traffic situations at the three-way intersection on the Perner Island in the city of Hallein, Austria (see Fig. 10.1). The investigated intersection was selected because of its mixed traffic that leads frequently to conflicting situations between vehicles, trucks and vulnerable road users like pedestrians or bicycle riders.

A dynamic management of traffic flow [16] including platoons needs not only a precise recognition of the current traffic situation but also a representative statistic over the real traffic situation over the long term in order to optimise for efficiency. This includes the meaningful aggregation of trajectories and the automatic identification of

T. Hoch (✉) · T. Kopetzky
Software Competence Center Hagenberg GmbH, Hagenberg, Austria
e-mail: thomas.hoch@scch.at

© The Author(s) 2022
A. Schirrer et al. (eds.), *Energy-Efficient and Semi-automated Truck Platooning*,
Lecture Notes in Intelligent Transportation and Infrastructure,
https://doi.org/10.1007/978-3-030-88682-0_10

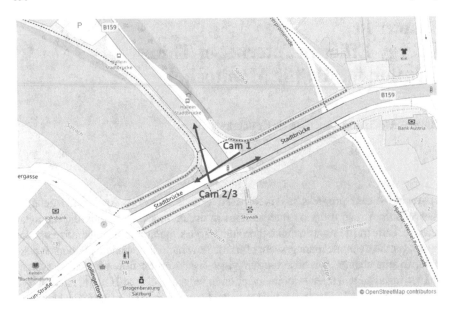

Fig. 10.1 Three-way intersection in Hallein where the traffic monitoring system was installed. The red arrows show the direction of view for each camera, base map and map data from OpenStreetMap, © OpenStreetMap contributors under the CC-BY-SA license, https://www.openstreetmap.org/copyright

the flow patterns at the intersection. We therefore developed in cooperation with the project partner *SWARCO FUTURIT Verkehrssignalsysteme GmbH* a video-based traffic measurement system that is able to locate all the traffic participants on the intersection and that can aggregate that information to reveal traffic flow patterns of different road users in great detail. We installed our system at the three-way intersection on the Perner Island in Hallein (see again Fig. 10.1) and tracked six different classes of road users for two weeks in order to get a clear understanding about the real traffic situations and flow patterns on the intersection.

Cameras are widely used for traffic monitoring at intersections and provide rich visual information about road users. A typical set-up consists of a single camera high mounted at a close by building such that the full crossing is observable [6, 24]. This set-up has the advantage that occlusions of road users through other road users are minimised due to the elevated view point and that the configuration process is simplified because the stitching and calibration of different views are not necessary. However, the installation of such a system can become cumbersome since the mounting of the system on a nearby building involves the house owner in the installation process or is not possible at all if the crossing is in a rural area. We therefore follow a different approach where we construct self-contained recognition units that can be easily attached to any traffic light system.

In addition, we were also looking for a more sustainable set-up in terms of energy consumption. There is a growing concern that the strong increase of energy consumption of the IT infrastructure, with the application of deep learning methods as one of the key drivers of this development, at this pace is not sustainable [22]. A currently widely used set-up for street monitoring cameras is to broadcast the video to the cloud and to do the processing of the video stream there. However, the transportation of the high-resolution video data is energy intensive [5] and also needs an Internet connection with high bandwidth in order to not risk processing instabilities. In the Connecting Austria project, we took a different approach taking the energy consumption of the whole system from the beginning of the design process into account. We designed our recognition units as edge computing [26] devices that are able to process the video stream in real time due to a dedicated low energy hardware accelerator for neural networks. Similar solutions have been recently proposed by [2, 17]. Different to their approach, we go one step further and process and analyse the behaviour of the road users in terms of real-world coordinates which allows us to map the extracted traffic patterns to a digital twin of the intersection.

In the following, we describe the developed traffic estimation system that is able to collect comprehensive information about the traffic situation in real time to estimate traffic density and flows of cars and trucks with high precision.

10.2   Low Energy Internet of Things Traffic Monitoring System

Our project aim was the development of a low energy camera-based traffic monitoring system that is able to recognise six different types of road users in real time. The designed system should be easy to deploy on existing traffic light installations and should be able to send the precise location of the road users at the intersection to an operator or to an back-end solutions that can use this information to initiate further actions like warning a driver.

Figure 10.2 shows our measurement set-up at the intersection where we installed one recognition unit for each arm of the three-way intersection. Every recognition units consists of a camera that is connected to a processing unit with a dedicated hardware accelerator for artificial intelligence applications. To preserve the privacy of the road users, the processing of the camera stream is done locally via the attached AI processing unit. Therefore, possible sensitive information never leaves the device, and only the object class and its position in the image are sent to our cloud server. As an information broker, we use a Kafka server,[1] which is a an open-source distributed event streaming platform often used in Internet of Things (IoT) scenarios. At our cloud server, the final processing includes three steps: first, the integration of each camera view into one common world view, second, the tracking of the objects over time, and third, the estimation of traffic flow according to our flow graph of the street crossing. Figure 10.3 shows exemplary the intermediate results of our processing pipeline. All processing steps are detailed in the following sections.

---

[1] https://kafka.apache.org/.

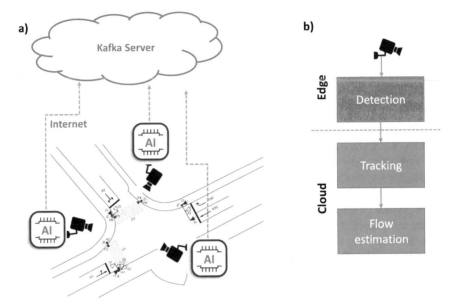

Fig. 10.2 **a** Measurement set-up at the three-way intersection in Hallein. **b** The processing pipeline. Object detection is done with a specialised AI processing unit attached to each of the three cameras. The detection results are transferred via a mobile Internet connection to the cloud infrastructure where the flow estimation is processed

10.2.1 Real-Time Object Detection

The last years showed tremendous progress in the field of object detection. One key driver was the development of new methods and tools that can be efficiently calculated on modern graphic cards. The now standard models like faster RCNN [19] or mask RCNN [8] achieve high accuracy but also need powerful server solutions for the processing of live video streams. Therefore, lightweight models like Yolo [18] or MobileNet [10] have been developed for the use on smartphones or embedded devices. Our system utilises an advanced architecture [25] derived from MobileNet that can be efficiently evaluated on the coral board[2] that we use to process the live stream of the cameras.

The coral board is an ARM-based single board computer with an on-board Edge TPU co-processor to perform fast machine learning (ML) inferencing. Although the board comes with an object detector, its performance for our setting is rather poor because of the limited images size of 300×300 pixels. In order to increase the detection performance of our system, we increased the image size to 960×384 pixels and trained a new model using a selection of images from the COCO data set [14] and images that we collected at the crossing. The final training data set

---

[2] https://coral.ai/.

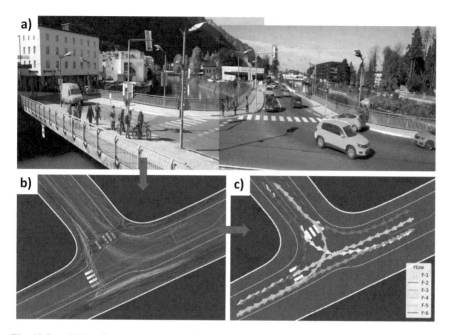

Fig. 10.3 a Object detection on camera images with 15 fps where we extract only the bounding box of objects. **b** The detected objects of each camera are transferred to lat/long world coordinates and via Kalman filtering we integrate them to object trajectories. **c** In a final step, the car trajectories are used to automatically infer the major traffic flows on the three-way intersection. In this case, six major traffic flows F-1 to F-6 are identified and depicted in different colours

contained approximately 20k images from the COCO data set and 10k images from the crossing in Hallein that we automatically annotated using the consensus estimate of two state-of-the-art networks [19, 23]. The final model was able to process images of size 960×384 with 15 frames per second (fps).

Several models were trained with the TensorFlow object detection toolkit[3] and quantised to 8 bit for the usage on the Coral device. The final model had a mean average precision of 82.6% on an manually labelled holdout data set consisting of 50 images from each of the three cameras at the crossing.

10.2.2   Sensor Fusion and Object Tracking

One important step in the configuration of our system is the projection of the image positions of recognised objects to the coordinate system of the crossing and to combine the projections of each camera to a common world view. Several techniques for roadside camera calibration are available [11] with the overconstrained approaches

---

[3] https://github.com/tensorflow/models/tree/master/research/object_detection.

Table 10.1 Class-specific relative position of the reference point in regard to the detection's bounding box

Class	Person	Car	Bicycle	Motorcycle	Bus	Truck
f_x	0.28	0.41	0.41	0.44	0.54	0.56
f_y	1.0	0.78	0.89	0.85	0.82	0.83

usually performing best. An accurate camera calibration facilitates the fusion of the projected camera positions and thus the object tracking as a whole. The final sensor fusion and tracking pipeline was implemented the following way.

First we calculated a class-specific reference point P_{ref} from the bounding box of every object detection as shown in the following equation

$$P_{ref}(x, y) = \left(x_1 + f_x \cdot (x_2 - x_1),\ y_1 + f_y \cdot (y_2 - y_1)\right),\qquad (10.1)$$

where x_1, y_1 are the top left coordinates and x_2, y_2 the right bottom coordinates of the bounding box, respectively. The factor f takes a value between $[0-1]$ and was optimised in such a way that the distance between the projected world coordinates of the same object seen from different camera views becomes a minimum. The derived values for f_x and f_y are shown in Table 10.1.

Second, with the help of the camera calibration toolbox from the opencv[4] library, we projected the image coordinate of the reference point to our world coordinate system. Each camera was calibrated manually using the visible subset of 20 carefully selected and mapped reference points on the intersection. Third, the projected world coordinates are then assigned to the predicted positions of tracked objects. For every tracked object, there can be maximal one assigned detection per camera. We used the Hungarian algorithm [12] to calculate an assignment with minimal distance between detections and tracked objects. For the Kalman filter update, we used the average position of the assigned detections. The Kalman filter was initialised with the discrete constant white noise kinetic model [4] that we parameterised for every object class individually.

10.2.3 Traffic Flow Estimation

In order to get a better understanding of the vehicle flows on the observed intersection, we used trajectory clustering [3] to group similar trajectories together and to automatically learn the possible traffic patterns at the intersection using an graph-based approach for traffic flow extraction [9]. The method first uses all trajectories to build a flow graph and then extracts flow patterns based on the maximum flow between to nodes of the graph. The major traffic flows of vehicles on the three-way

---

[4] https://opencv.org/.

Table 10.2 Evaluation result of the traffic flow estimation

Class	Hour	True count	Est. count	Flow combined		Flow pattern	
				Deviation	Error (%)	Deviation	Error (%)
Car	12–13	1244	1286	42	3.4	138	11.1
	14–15	1186	1134	52	4.4	124	10.5
Truck	12–13	72	64	8	11.1	20	27.8
	14–15	56	54	2	3.6	14	25.0

The table shows the error in the count estimate averaged over all six traffic flow patterns combined and separated

intersection are depicted in Fig. 10.3c. The measured vehicle trajectories are mapped to one of the six paths based on the minimal average distance between trajectory points and flow path. Because of tracking errors due to object occlusions or alignment failures, a significant number of tracks could be only observed partly and thus had to be removed from the analysis in order to get a good estimate of the vehicle counts. This was done with an additional calibration step that excluded short tracks due to tracking problems. We used two one-hour recordings with ground truth of the car and truck count to adapt the counting method parameters and measure the flow estimation quality in a cross-validation setting.

10.3   Traffic Flow Measurement Result

First we evaluate our traffic flow estimation method with manually obtained car and truck counts at the three-way intersection. Table 10.2 summarises the true and the estimated count of cars and trucks over two one-hour observation periods. For every observation period, we used the other one to perform a hyper-parameter optimisation of our estimation method. The measured deviation from the true count averaged over the two observation periods was 3.9% for cars and 7.4% for trucks, respectively. Second we also evaluated the precision of the object count for the six observed traffic flow patterns at the intersection (see Fig. 10.3c) individually. The average error of the count estimate increased to 10.8% for cars and to 26.4% for trucks. The strong increase of the error for the truck class is a result of the general low frequency of trucks for flows F-3 to F-6 which makes a precise estimation of the traffic flow more unreliable.

Furthermore, we investigated the traffic flow at the intersection for a two weeks observation period. Figure 10.4 shows the estimate flow of cars and trucks over a 12 h time span on an hourly basis. The measurement was done from 18 September till 1 October 2020. The estimates are calculated separately for workdays (orange bars) and weekends (blue bars) since the amount of truck traffic changes considerable during weekends. Whereas the truck traffic has its peak during the morning hours and decays considerably in the afternoon and evening hours, the car traffic stays

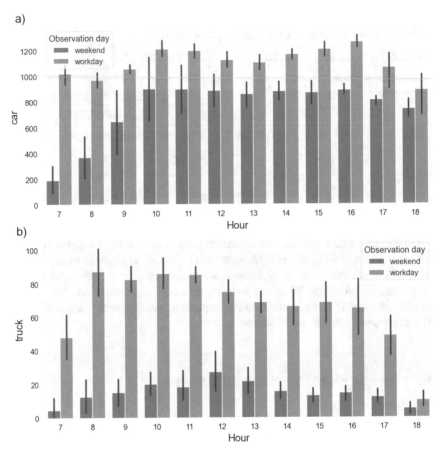

Fig. 10.4 Object count estimates of cars (**a**) and trucks (**b**) from 7am to 7pm on an hourly basis. The bars show the mean count of the object class per hour, and the black line indicates the variation of the count estimate. Orange bars are averaged over workdays, whereas the blue bars show the average of weekends

almost constant and decreases only in the evening. The average percentage of trucks on the intersection during this measurement period was 6.0%. During weekends, the average percentage of trucks decreased to 1.8% due to a considerable lower number of trucks passing the intersection (Fig. 10.4b). For cars on the other hand, we find a strong decrease in the count only in the morning hours due to Sundays.

We also investigated how the car and truck traffic is distributed along the six traffic flow patterns. Figure 10.5 shows that most traffic is along flow pattern F-1 and F-2 (see Fig. 10.3c for the definition of the flow patterns) which is a higher ranked street that bypasses the old town. F-3 to F-6 are distributor roads from and into the old town that show less traffic (also because of a smaller time share on the traffic light switching schedule). It can be clearly seen that the truck traffic from and into the

Table 10.3 Percentage of trucks on the intersection during weekdays partitioned by traffic flow patterns

Quantile	F-1	F-2	F-3	F-4	F-5	F-6
0.25	5.4	3.8	1.3	5.0	0.0	0.0
0.75	9.2	7.2	3.7	19.3	3.7	0.0

old town is very low and thus also the percentage of trucks on the intersection as Table 10.3 shows. The only exception is the distributional road from the old town (traffic pattern F-4) for which we observed the highest variation in the percentage of trucks on the intersection. Although the number of cars and trucks along this flow pattern is generally low and leads to high variation in the estimate, Fig. 10.5b also shows that there is significant truck traffic along this flow pattern that explains the strong increase in the percentage of trucks on the intersection.

10.4   Discussion

The evaluation measurement was done on a sunny day with very similar weather conditions between the two observation periods. It is well known that video-based tracking systems are sensitive to weather conditions like fog or snow [6], and we expect an increase in error for such conditions. Simulation of weather conditions via style transfers as in [15] could in principle help to generate a more precise evaluation of the system. However, in our case it was not possible to run an additional deep learning model on the embedded board due to a limited memory and processing capacity.

One difficulty we observed during the execution of this study was the classification of vehicles into separate car and truck classes. During evaluation, we found a gradual transition from car over van to truck class where it was not always easy to draw a clear border between these classes based on vehicle features. Furthermore, our vehicle classification model produced a rather coarse separation of vehicle classes due to the use of pretrained models for the automatically annotation procedure that are aimed for a more general recognition task. Therefore, to get a more standardized vehicle classification as outlined in [7], it would be necessary to extend the training set generator with a specialised classification network as described in MATLAB® [21].

A key feature of our solution was the simple installation procedure on the traffic light itself. The developed device is self-contained with a build in mobile connection to our cloud service and thus needs no wired connection to the Internet which is usually not available at intersections. Although this gives more flexibility in positioning of the device at the traffic light, we also observed that the smaller distance to the road leads to more occlusion of cars and other road users by big vehicles like trucks and buses. Because of these occlusions, a substantial number of vehicles could be tracked only partly leading to more than one track per vehicle. Another factor contributing

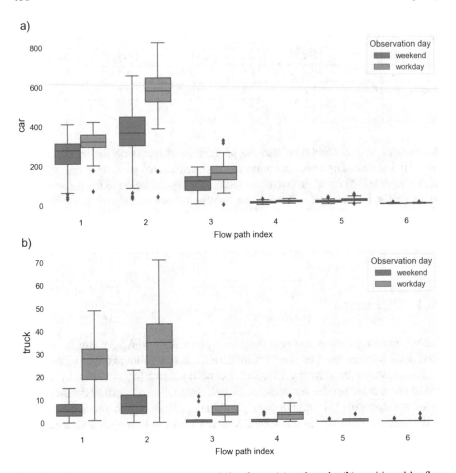

Fig. 10.5 Figure shows the object count statistic of cars (**a**) and trucks (**b**) partitioned by flow patterns as defined in Fig. 10.3c. Orange box plots show the summary statistic for workdays, whereas the blue box plots show for weekends, respectively

to this problem is the difficulty of calculating the correct position of vehicles that are only partly seen in the video. In this case, the reference point calculation as described in Sect. 10.2.2 leads to an offset in the position estimate that makes the prolongation of trajectories between camera views more error prone. To circumvent this issue, we had to carefully tune the hyper-parameters of the trajectory selection process. We also observed a higher variation in object size of vehicles that made their recognition more difficult. Since our units were positioned at the roadside, we expect that a more central arrangement above the road could bring advantages.

One specific goal of this investigation was to build a low energy recognition system. Deep learning algorithms are energy hungry [13] and contribute significantly to the increase of energy consumption of the IT infrastructure [22]. Therefore, a sustainable traffic monitoring solution needs to take the energy consumption of the system into account since a nationwide enrolment would mean the installation of thousands of devices. The presented solution is based on the Coral Edge TPU which provides an energy-efficient way for object detection. The average power used by the device for the video stream processing was approximately 4.9 W (2.4 W in idle mode). Thus with the combination of our technology stack with newly proposed methods for low energy communication in 5G networks [1], an energy-efficient traffic monitoring platform is feasible.

10.5   Conclusion and Outlook

In this chapter, we presented a modern traffic measurement system that has four key advantages over conventional systems: (1) low energy consumption due to edge computing, (2) distributed logic edge and cloud results in a cost-efficient solution, (3) local processing grants a high level of privacy and (4) self-contained field device supports easy on-site installation.

We demonstrated that the system is able to measure the traffic flow of cars and trucks at the three-way intersection in Hallein with high precision and that we are able to partition the vehicle flow into one of the six automatically extracted flow patterns. Our analysis gives more insight on the spatial and temporal distribution of the car and truck traffic at the intersection and provides a basis for a more detailed scenario-based simulation approach in the Connecting Austria project.

In this work, we focused solely on traffic flow measurement of vehicles. The described measurement system can be also used to track and analyse the behaviour of vulnerable road users at an urban intersection. The precise location of these road users could be used to generate C-ITS messages that warn approaching vehicles of potentially dangerous situations as for example "person on the road". For such a use case, it is critical that the necessary information is provided within a short time frame. Although the measurement frequency of our system with 15 frames per second would in principle allow for such a fast processing, the current design is not favourable for this use case, since the communication with the cloud introduces some significant delays with traditional communication networks. Thus, the latency of such a system is a key factor which we will consider in the future development of our system.

References

1. Al Homssi B, Al-Hourani A, Chavez KG, Chandrasekharan S, Kandeepan S, Energy-efficient IoT for 5G: a framework for adaptive power and rate control. Technical report
2. Barthélemy J, Verstaevel N, Forehead H, Perez P (2019) Edge-computing video analytics for real-time traffic monitoring in a smart city. Sensors (Switzerland) 19(9):5
3. Bian J, Tian D, Tang Y, Tao D (2018) A survey on trajectory clustering analysis 2
4. Blackman SS, Popoli R (1999) Design and analysis of modern tracking systems

5. Bolla R, Bruschi R, Davoli F, Cucchietti F (2011) Energy efficiency in the future internet: a survey of existing approaches and trends in energy-aware fixed network infrastructures
6. Datondji SRE, Dupuis Y, Subirats P, Vasseur P (2016) A survey of vision-based traffic monitoring of road intersections 10
7. Hallenbeck M, Selezneva O, Quinley R (2014) Verification, refinement, and applicability of long-term pavement performance vehicle classification rules. Technical report
8. He K, Gkioxari G, Dollár P, Girshick R (2017) Mask R-CNN, 3
9. Hoch T (2021) A spatial knowledge graph for human behavior pattern extraction (in preparation)
10. Howard AG, Zhu M, Chen B, Kalenichenko D, Wang W, Weyand T, Andreetto M, Adam H (2017) MobileNets: efficient convolutional neural networks for mobile vision applications, 4
11. Kanhere NK, Birchfield ST (2010) A taxonomy and analysis of camera calibration methods for traffic monitoring applications. Technical report
12. Kuhn HW (1955) The Hungarian method for the assignment problem. Naval Res Logistics Q 2(1–2)
13. Li D, Chen X, Becchi M, Zong Z (2016) Evaluating the energy efficiency of deep convolutional neural networks on CPUs and GPUs. In: Proceedings—2016 IEEE international conferences on big data and cloud computing, social computing and networking, and sustainable computing and communications. Institute of Electrical and Electronics Engineers Inc., 10, pages 477–484
14. Lin T-Y, Maire M, Belongie S, Bourdev L, Girshick R, Hays J, Perona P, Ramanan D, Lawrence Zitnick C, Dollár P (2014) Microsoft COCO: common objects in context, 5
15. Luan F, Paris S, Shechtman E, Bala K (2017) Deep photo style transfer 3
16. Nellore K, Hancke GP (2016) A survey on urban traffic management system using wireless sensor networks, 1
17. Nikodem M, Słabicki M, Surmacz T, Mrówka P, Dołęga C (2020) Multi-camera vehicle tracking using edge computing and low-power communication. Sensors (Switzerland) 20(11):1–16
18. Redmon J, Divvala S, Girshick R, Farhadi A (2015) You only look once: unified, real-time object detection
19. Ren S, He K, Girshick R, Sun J (2015) Faster R-CNN: towards real-time object detection with region proposal networks
20. Schildorfer W, Kuhn A, Walter A (2019) Connecting Austria-first results of C-ITS-focused level 1 truck platooning deployment HiTec-an independent non-for-profit research institution into innovation. Technical report
21. Sun W, Zhang X, Shi S, He X (2019) Vehicle classification approach based on the combined texture and shape features with a compressive DL. IET Intell Transp Syst 13(7)
22. Vinuesa R, Azizpour H, Leite I, Balaam M, Dignum V, Domisch S, Felländer A, Daniela Langhans S, Tegmark M, Fuso Nerini F (2020) The role of artificial intelligence in achieving the sustainable development goals, 12
23. Wang J, Sun K, Cheng T, Jiang B, Deng C, Zhao Y, Liu D, Mu Y, Tan M, Wang X, Liu W, Xiao B (2020) Deep high-resolution representation learning for visual recognition. IEEE Trans Pattern Anal Mach Intell
24. Wang X (2013) Intelligent multi-camera video surveillance: a review. Pattern Recogn Lett 34(1):3–19
25. Xiong Y, Liu H, Gupta S, Akin B, Bender G, Kindermans P-J, Tan M, Singh V, Chen B (2020) MobileDets: searching for object detection architectures for mobile accelerators, 4
26. Zhou Z, Chen X, Li E, Zeng L, Luo K, Zhang J (2019) Edge intelligence: paving the last mile of artificial intelligence with edge computing. Proc IEEE

Thomas Hoch is a key researcher at the Software Competence Center Hagenberg GmbH for data science. His research interests cover the area of computer vision, scene understanding as well as human behaviour modelling. Thomas Hoch has been engaged in many national and international projects as a principle investigator as for example in the EU projects Tresspass and Teaming.AI.

Theodorich Kopetzky is area manager for Services and Solutions at the Software Competence Center Hagenberg GmbH since the beginning of 2020. Before that he was executive head of the "Knowledge-based Vision Systems" research focus. He has been engaged in many national and international projects, particularly noteworthy as project manager for a multi person-year project in the area of SOA based enterprise application platform systems.

Chapter 11
Fuel Efficiency Assessment

José Carmona, David Hildenbrandt, Florian Hofbauer,
and Matthias Neubauer

Abstract The assessment of fuel efficiency represents a vital element when it comes to the deployment and business model development of truck platooning. In this chapter, the methodological approach implemented in the Connecting Austria project to assess fuel efficiency is presented. The approach covers the following assessment aspects: (1) the assessment of the road infrastructure in terms of the suitability of road segments for truck platooning, (2) the assessment of driving behaviour and strategies for truck platoon formation and dissolution and (3) the assessment of efficiency in terms fuel savings for certain routes.

Keywords Road infrastructure assessment · Driving behaviour assessment · Efficiency assessment

11.1 Road Infrastructure Assessment

The road infrastructure assessment represents the initial building block of the proposed fuel efficiency assessment methodology. Figure 11.1 visualises the overall assessment approach and specifies required input as well as output data; e.g. the assessment of the road infrastructure (i) requires data on given platooning restrictions as well as the underlying road network and (ii) generates a risk-rated road network as output. The assessment aspects are detailed in the subsequent sections. As reported in Chap. 3, previous studies already investigated potential opportunities and risks of truck platooning as well as required technical and infrastructural conditions. The assessment of the road infrastructure with respect to specific truck platooning regulations represents a vital starting point for subsequently assessing fuel efficiency in certain regulation scenarios. A list of truck platooning regulation

J. Carmona · D. Hildenbrandt
Andata Entwicklungstechnologie GmbH, Hallein, Austria

F. Hofbauer · M. Neubauer (✉)
Department of Logistics, University of Applied Sciences Upper Austria, Steyr, Austria
e-mail: matthias.neubauer@fh-steyr.at

© The Author(s) 2022
A. Schirrer et al. (eds.), *Energy-Efficient and Semi-automated Truck Platooning*,
Lecture Notes in Intelligent Transportation and Infrastructure,
https://doi.org/10.1007/978-3-030-88682-0_11

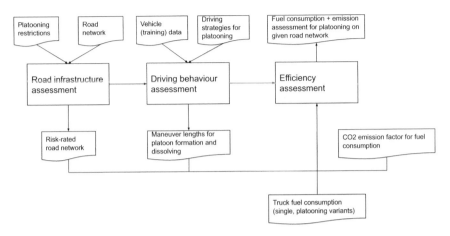

Fig. 11.1 Fuel efficiency assessment methodology—overview

scenarios is presented in Fig. 3 in Chap. 3 based on related work and results gained by the Austrian council for road safety.

In the Connecting Austria project, a software solution has been developed, which allows to generate a "risk-rated map" for a platooning regulation scenario comprising different constraints, e.g. restriction of platooning in tunnels or on bridges. The software solution allows to asses different scenarios and generates a road graph consisting of platooning-eligible and non-eligible road segments for a defined route. A visualisation of an example result is depicted in Fig. 11.2. In the following section, the software application that was used to generate the results presented in Chap. 12 will be discussed in further detail.

11.1.1 Risk-Rated Map

A given road network may consist of several possible zones of higher risk that are restricted for platooning. Given the assumption that a platoon needs to extend the distance between vehicles in a platoon and fully dissolve before it reaches an area that is restricted for platooning, one might ask the question "How many net kilometres will be left for platooning once all potential locations of higher risk and tolerance areas around them are excluded?". In order to answer this question with respect to underlying assumptions, ANDATA Artificial Intelligence Labs[1] developed a solution strategy and implemented a software application. In the following, the assumptions, the solution strategy and a test example for the generation of a risk-rated map for the Austrian motorway are sketched.

---

[1] https://www.andata.at/en/home.html.

Table 11.1 Input file parameters of test data set D

Parameter	Bridge	Entry	Exit	Tunnel
Tolerance before (m)	1000	2000	2000	1000
Tolerance after (m)	1000	2000	2000	1000
Minimal length (m)	100	–	–	–

Assumptions

- Vehicles in a platoon are obliged to extend the intra-vehicle distance to the normal safety distance ($\geq$ 50m intra-vehicle distance) before reaching a zone of higher risk that is restricted for platooning.
- Segments that are non-eligible for platooning can be categorised into:
  - Static higher risk segments, e.g. entries, exits, bridges, tunnels.
  - Quasi-static higher risk segments, e.g. long-term building sites.
  - Dynamic higher risk segments, e.g. road traffic accidents and patch of fog.

Solution Strategy

The road infrastructure assessment depends on the given road network and the defined platooning restrictions. Since the restrictions might be modified with regard to different platooning situations, a generic solution supporting a flexible configuration of restrictions based on an input file has been developed. The implemented software application may generate a (colour-coded) road segment classification depending on an input file. Doing so, routes may be analysed, e.g. with respect to the amount of platooning-eligible kilometres or restricted road segments in a given configuration.

The application generates a data processing pipeline based on available geographic information databases (e.g. GIP and OpenStreetMap), which calculates all possible routes that remain free for platooning, depending on the parameters defined in the input configuration file. The underlying architecture of this data processing pipeline is flexible in the sense that dynamically changing streams of data (e.g. traffic/weather information) could be fed into it. The outputs of the application are:

- A map showing all road segments specified in the input configuration file, where areas that are eligible for platooning are labelled green and areas that are unsafe for platooning are labelled red (risk-rated map).
- Statistical data, e.g. the total number of zones that are deemed safe for platooning and the distribution of lengths of distances between platooning and non-platooning zones.

Test Data Set D

In order to demonstrate the functionality of the application, a set of parameters listed in Table 11.1 was tested (matching the "play safe" scenario in Chap. 3) for the motorway A1.

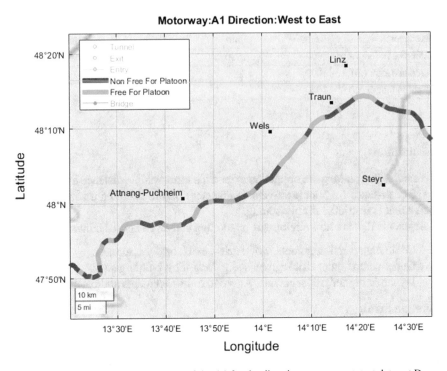

Fig. 11.2 Risk-rated map of a segment of the A1 for the direction west to east: test data set D

A segment of the resulting risk-rated map of the motorway A1 in direction west to east is depicted in Fig. 11.2. Resulting characteristic numbers, e.g. the number of kilometres available for platooning given the input parameters, are listed in Table 11.2. A more detailed view of the distribution of lengths of segments that are eligible for platooning is displayed in the form of an empirical cumulative distribution function plot in Fig. 11.3 on the left-hand side. On the right-hand side, the distribution of the distances between these segments is depicted. According to Fig. 11.3, the maximum length of road segments that are available for platooning is below 8 km and about 90% of all segments eligible for platooning are below 6 km of length for this input parameter configuration.

11.2 Driving Behaviour Assessment

The formation and dissolution of truck platoons may build upon different driving behaviours and strategies. Within the Connecting Austria project studies, different strategies for forming and dissolving truck platoons were defined and simulated on a microscopic level.

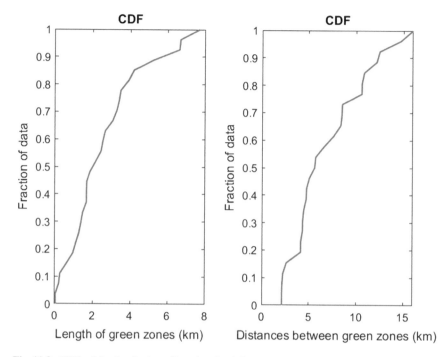

Fig. 11.3 CDFs of the distribution of lengths of and distances between segments that are eligible for platooning on the motorway A1 in direction from west to east for the input parameters of scenario D

Table 11.2 Results for test set D on the motorway A1 for the direction west to east

Characteristic number	Value
Total number of kilometres	292.4
Total number of kilometres available for platooning	71.86
Total number of available areas for platooning	27
Total number of non-available areas for platooning	28
Percentage of area available for platooning	24.57 %

Regarding platoon formation, a collaborative strategy was defined, where all of the trucks adapt their speeds in order to close the gap between them and drive as a platoon. In general, the compression of the platoon occurs primarily by braking and accelerating the front platoon vehicles. Figure 11.4 shows some possible trajectories for solving the platoon formation problem.

Regarding ad hoc platoon formation, that is, cases where there is an already formed platoon, and a truck that intends to join the platoon, the following possibilities were considered:

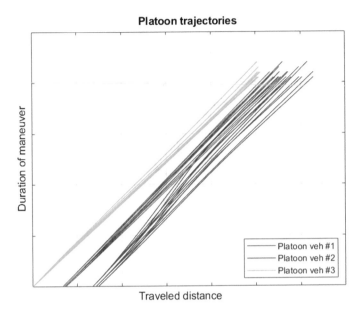

Fig. 11.4 Example of trajectories for platoon compression

- The truck is behind the platoon. The platoon does not act collaboratively; the truck has to accelerate in order to reach the platoon.
- The truck drives in front of the platoon. The platoon does not act collaboratively; the truck has to reduce its speed in order to reach the platoon.
- The platoon and the free truck collaborate, each one of them adapting the speed in order for the truck to join the platoon.

Regarding platoon dissolution, a collaborative strategy was defined, where all of the platoon trucks adapt their speeds. At the end of the dissolution manoeuvres, it is assumed that the trucks drive at their original speeds.

The fuel consumption for manoeuvres was calculated based on the vehicle speed, the intra-platoon distance and the formation/dissolution strategy. The additional fuel consumption was clustered according to intra-platoon distance, manoeuvre lengths for formation and dissolution manoeuvres, and the dedicated trucks in a platoon. Clusters for manoeuvre lengths were defined as follows:

- 0–499m manoeuvre length → fast formation/dissolution strategy.
- 500–1199m manoeuvre length → medium formation/dissolution strategy.
- 1200–2500m manoeuvre length → slow formation/dissolution strategy.

In order to derive fuel consumption models, a database of several 10000 truck kilometres driven on Austrian motorways and the dedicated fuel consumption has been evaluated. A visualisation of the distribution of truck routes in a typical data set within this database is shown in Fig. 11.5.

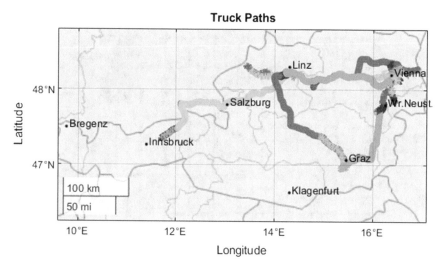

Fig. 11.5 Visualisation of truck routes in a typical data set. The colours highlight several of the longer continuous paths contained within the data set

In the course of obtaining suitable training data sets for a fuel consumption prediction model, data preprocessing routines were implemented. Trajectories which proved to be unsuitable for the purpose of predicting fuel consumption were excluded. For example, a filter that yielded only trajectories on motorways inside a speed range between 60 and 90 km/h was applied. Furthermore, data sets with faulty or implausible speed sensor readings, e.g. instant velocity drops or surges outside a certain boundary, were excluded. A machine learning model was then calibrated to predict the fuel consumption of a typical truck inside the speed range of interest. By means of an elaborated feature engineering process, the most suitable parameters and combinations thereof were derived for the prediction of the total fuel consumption on a given trajectory.

The results for additional fuel consumption for the formation and dissolution of a truck platoon depending on the intra-platoon distances 0.5, 1, 1.5 and 2s and the different driving behaviours (fast, medium, slow) are summarised in Tables 11.3 and 11.4 and visualised in Fig. 11.6. These results serve as basis to assess fuel efficiency and economic viability (see Sect. 11.3) of transport routes.

11.3 Efficiency Assessment

In this section, the efficiency assessment approach developed and applied in the Connecting Austria project is described. Relevant input data for the assessment represents (compare Fig. 11.1):

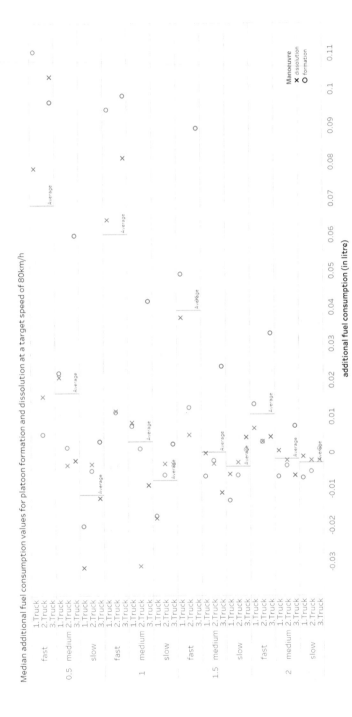

Fig. 11.6 Median values for additional fuel consumption at intra-platoon distances 0.5, 1, 1.5, 2s

Table 11.3 Median fuel consumption and manoeuvre length for platoon formation and dissolution at a target speed of 80 km/h (1)

Intra-platoon distance (s)	0.5			1		
Driving behaviour (strategy)	Fast	Medium	Slow	Fast	Medium	Slow
Platoon formation						
Manoeuvre length (km)	0.497	0.863	1.848	0.494	0.850	1.825
1.Truck (+consumption in litre)	0.1105	0.0222	−0.02	0.0946	0.0075	−0.0172
2. Truck(+consumption in litre)	0.0052	0.0017	−0.0048	0.0113	0.0013	−0.006
3. Truck(+consumption in litre)	0.0967	0.06	0.0033	0.0984	0.0419	0.0026
Platoon dissolution						
Manoeuvre length (km)	0.433	0.842	1.631	0.371	0.825	1.653
1. Truck (+consumption in litre)	0.0785	0.0209	−0.0315	0.0643	0.0085	−0.0179
2. Truck (+consumption in litre)	0.0156	−0.0033	−0.003	0.0116	−0.031	−0.0029
3. Truck (+consumption in litre)	0.1037	−0.002	−0.0124	0.0813	−0.0088	−0.003

Table 11.4 Median fuel consumption and manoeuvre length for platoon formation and dissolution at a target speed of 80 km/h (2)

Intra-platoon distance (s)	1.5			2		
Driving behaviour (strategy)	Fast	Medium	Slow	Fast	Medium	Slow
Platoon formation						
Manoeuvre length (km)	0.474	0.859	1.824	0.459	0.835	1.801
1. Truck (+consumption in litre)	0.0493	−0.0063	−0.013	0.0135	−0.0064	−0.0068
2. Truck (+consumption in litre)	0.0126	−0.002	−0.0061	0.0032	−0.0034	−0.005
3. Truck (+consumption in litre)	0.0895	0.0239	0.001	0.033	0.0075	0.0012
Platoon dissolution						
Manoeuvre length (km)	0.376	0.831	1.670	0.382	0.827	1.626
1. Truck (+consumption in litre)	0.0373	0	−0.0058	0.0069	0.0007	−0.0009
2. Truck (+consumption in litre)	0.0051	−0.0029	−0.0025	0.0032	−0.002	−0.002
3. Truck (+consumption in litre)	0.0425	−0.0109	0.0044	0.0045	−0.0061	−0.002

- **Risk-rated road network**—generated on defined scenarios for a certain route, comprising information on platooning ability of road segments.
- **Manoeuvre lengths for platoon formation and dissolving**—depending on different strategies to form and dissolve a platoon.
- **Fuel consumption models**—for solo trucks and trucks in a platoon, depending on intra-platoon distance and speed.
- **CO_2 emission factor for fuel consumption**—to estimate CO_2 emissions.

Based on the risk-rated road network, net platooning distances may be calculated as illustrated in Fig. 11.7 and formula 11.1 for each road segment in the network. The assessment may consider two aspects:

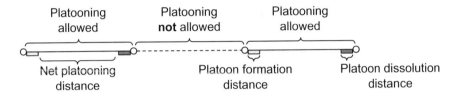

Fig. 11.7 Net platooning distance illustration

1. The assessment of the general platooning feasibility for a given route.
2. The assessment of economically viable platooning operations for a given route.

The two aspects are detailed within Sect. 11.3.1 and 11.3.2.

$$d_{\text{net\_platooning}} = d_{\text{platooning\_allowed}} - d_{\text{platoon\_formation}} - d_{\text{platoon\_dissolution}} \qquad (11.1)$$

11.3.1 General Feasibility of Platooning on a Road Segment

The assessment of the general feasibility of platooning on a road segment is detailed in Algorithm 11.1. Basically, road segments where platooning is allowed and the length of the road segment is greater than the distance required to form and dissolve a platoon are accumulated to the net platoon length of a given route. This net platooning length may serve as basis to assess fuel and emission savings based on fuel efficiency effects of truck platooning. However, costs for forming and dissolving a platoon may decrease saving potentials. Therefore, a further investigation of economically viable net platooning distances is considered appropriate (cf. Sect. 11.3.2).

11.3.2 Economic Viability of Platooning on a Road Segment

The economic viability takes into account additional fuel consumption for platoon formation and dissolution. To assess these additional costs, reference values from microscopic traffic simulations based on diverse platoon formation and dissolution strategies (cf. Sect. 11.2) may be used. The approach for this assessment is detailed in Algorithms 11.2 and 11.3. In case the accumulated solo consumption of the number trucks in a platoon is greater than the consumption of the trucks driving in a platoon (including additional costs for formation and dissolution), a route segment is considered economically viable.

The resulting net platooning distance rated with a truck platooning fuel efficiency saving model that takes into account savings in relation to intra-platoon distances plus the additional fuel consumption for platoon formation and dissolution allow to

Algorithm 11.1 Calculation of net platooning and non-platooning distances

```
 1: Initialise l_road_segments for a specific route;
 2: Initialise d_netPlatooning;
 3: Initialise d_soloDrive;
 4: Initialise d_platoon_form_diss; // length for platoon formation and dissolution
 5:
 6: for r_seg IN l_road_segments do
 7:     if r_seg.platooningIsAllowed == false then
 8:         d_soloDrive += r_seg.length;
 9:     else // Platooning is allowed on road segment
10:         if r_seg.length >= d_platoon_form_diss then
11:             d_netPlatooning += r_seg.length - d_platoon_form_diss;
12:             d_soloDrive += d_platoon_form_diss;
13:         else // road segment < length for formation and dissolution
14:             d_soloDrive += r_seg.length;
15:         end if
16:     end if
17: end for
```

Algorithm 11.2 Economic optimisation of net- and non-platooning distances

```
 1: Initialise l_road_segments for a specific route;
 2: Initialise d_netPlatooning;
 3: Initialise d_soloDrive;
 4: Initialise d_platoon_form_diss; // length for platoon formation and dissolution
 5:
 6: Initialise isEconomicallyViable;
 7: Initialise formationStrategy;
 8: Initialise dissolutionStrategy;
 9: Initialise intraPlatoonDist;
10: Initialise pFuelConsumptionModel;
11: Initialise soloConsumption;
12:
13: for r_seg IN l_road_segments do
14:
15: isEconomicallyViable = isEconomicallyViable(r_seg, formationStrategy, dissolutionStrat-
    egy, intraPlatoonDist, pFuelConsumptionModel, soloConsumption, nrOfTrucksInPlatoon,
    d_platoon_form_diss);
16:
17:     if r_seg.platooningIsAllowed == false then
18:         d_soloDrive += r_seg.length;
19:     else // Platooning is allowed on road segment
20:         if r_seg.length >= d_platoon_form_diss AND isEconomicallyViable then
21:             d_netPlatooning += r_seg.length - d_platoon_form_diss;
22:             d_soloDrive += d_platoon_form_diss;
23:         else // road segment < length for formation and dissolution
24:             d_soloDrive += r_seg.length;
25:         end if
26:     end if
27: end for
```

Algorithm 11.3 Economic viability assessment

1: Initialise r_seg;
2: Initialise formationStrategy;
3: Initialise dissolutionStrategy;
4: Initialise intraPlatoonDist;
5: Initialise pFuelConsumptionModel;
6: Initialise soloConsumption;
7: Initialise nrOfTrucksInPlatoon;
8: Initialise d_platoon_form_diss;
9:
10: **if** r_seg.platooningIsAllowed == true **AND** r_seg.length >= d_platoon_form_diss **then**
11:
12: soloConsumptionOfTrucks = r_seg.length * soloConsumption * nrOfTrucksInPlatoon;
13:
14: platooningConsumptionOfTrucks =
15:      fuelConsumptionFormation(formationStrategy) +
16:      fuelConsumptionDissolution(dissolutionStrategy) +
17:      fuelConsumptionNetPlatooning(pFuelConsumptionModel, intraPlatoonDist);
18:
19:     **if** soloConsumptionOfTrucks > platooningConsumptionOfTrucks **then**
20:         **return** true;
21:     **else**
22:         **return** false;
23:     **end if**
24: **else**
25:     **return** false;
26: **end if**

---

asses fuel efficiency for certain use cases. Applications of the described procedure will be presented in the following Chap. 12.

11.4 Conclusion

This chapter described the methodological approach taken in the Connecting Austria project to assess fuel efficiency. Doing so, the approaches for the three aspects road infrastructure, driving behaviour and efficiency were detailed. Within the assessment of the road infrastructure, the suitability of road segments for truck platooning is rated based on configurable platooning regulations that allow a scenario-based assessment of fuel efficiency considering diverse regulations. Furthermore, the assessment of driving behaviour and strategies for truck platoon formation and dissolution provided a required basis to evaluate the economic viability of platooning-eligible road segments (see Sect. 11.3). The resulting segments suitable for platooning serve as basis to assess fuel savings for certain use application (see Chap. 12).

José Carmona Ph.D. in Computer Science and Computer Science Engineer from the University of Málaga (Spain), is a development engineer at Andata GmbH. His research interest include computational intelligence, data mining and simulation in connection with traffic automation.

David Hildenbrandt has received a Masters Degree in Physics from the University of Vienna. He works as a development engineer and data scientist at ANDATA GmbH and his research interests include computational intelligence, stochastic and nonlinear dynamics as well as modelling and simulation of intelligent transportation systems.

Florian Hofbauer   is a research associate in the research group hyperconnected logistics systems at the University of Applied Sciences Upper Austria. His research focus is on sustainability assessment of intelligent transport systems, automated mobility, alternative fuels and sustainable transport systems.

Matthias Neubauer is a professor at the University of Applied Sciences Upper Austria for logistics information systems. His research interests cover human-computer interaction, intelligent transportation systems as well as cooperative, connected and automated mobility. Matthias is involved in international and national research projects and teaches in the masters program digital transport and logistics management classes such as process management, distributed logistics systems or geoinformation systems.

Chapter 12
Application of Fuel Efficiency and Traffic Efficiency Assessment

Elvira Thonhofer, Matthias Neubauer, and Florian Hofbauer

Abstract This chapter presents the application of the fuel assessment methodology developed in the Connecting Austria project. Thereby, a route analysis for an Austrian fleet operator is performed including the assessment of feasible and economic viable routes and scenarios. Furthermore, potential fuel consumption and CO_2 emission savings are discussed within the given case. The saving potential may be increased via dynamic C-ITS-based truck platoon regulations, instead of statically defined, too restrictive regulations as indicated in the C-ITS assessment section. Finally, the chapter discusses the effect of truck platooning on increasing traffic efficiency.

Keywords Fuel efficiency · Traffic efficiency · C-ITS (cooperative intelligent transport systems)

12.1 Fuel Efficiency Assessment in a Fleet Operator Case

The evaluation of fuel consumption and its sensitivity on selected parameters is carried out to show the underlying trade-offs with respect to platooning fuel efficiency. These selected parameters are:

- Risk-rated map configuration (how often does the platoon have to be created/dissolved (i.e. number of required formation/dissolution manoeuvres) on a given route.
- Swiftness of the platoon manoeuvres (highly active platoon vs. slow response).
- Intra-platoon distance (time gap between trucks).

The evaluation will take into account the methodology presented in Chap. 11 and investigate given routes of an Austrian fleet operator. Aligned with an Austrian

E. Thonhofer
Andata Entwicklungstechnologie GmbH, 5400 Hallein, Austria

M. Neubauer (✉) · F. Hofbauer
Department of Logistics, University of Applied Sciences Upper Austria, Steyr, Austria
e-mail: matthias.neubauer@fh-steyr.at

© The Author(s) 2022   157
A. Schirrer et al. (eds.), *Energy-Efficient and Semi-automated Truck Platooning*,
Lecture Notes in Intelligent Transportation and Infrastructure,
https://doi.org/10.1007/978-3-030-88682-0_12

fleet operator, typical inner Austrian routes on highways were defined. These routes have already been illustrated in Fig. 3.4. The routes comprise only Austrian highways, since the assessment focused on truck platooning potentials on highways. The following transport routes (back and forth) were selected:

- Pasching–Guntramsdorf (concerning the Austrian highways A1, A21).
- Pasching–Werndorf (concerning the Austrian highways A1, A9).
- Pasching–Kalsdorf (concerning the Austrian highways A1, A9).

In an initial step, a road infrastructure assessment (compare Fig. 11.1) was conducted. Thereby, a risk-rated map was developed to identify road segments that are eligible and non-eligible for platooning taking into account the defined truck platooning restriction scenarios (compare Fig. 3.3).

Given the resulting scenario-based risk-rated map data, the assessment of the feasibility and economic viability of the routes was conducted. The results for the direction Pasching–Guntramsdorf are depicted in Fig. 12.1. The figure highlights the results based on the scenarios A–F and data for an intra-platoon distance of 0.5 s at a target speed of 80 km/h. For each scenario, the following information is provided:

- **Platooning allowed**—number of road segments where platooning is allowed with respect to the scenario regulations.
- **Platooning feasible**—number of road segments where platooning is feasible with respect to the minimum length for platoon formation and dissolution.
- **Fast form/diss**—number of road segments that are economically viable in case of applying the fast formation and dissolution strategy.
- **Medium form/diss**—number of road segments that are economically viable in case of applying the medium formation and dissolution strategy.
- **Slow form/diss**—number of road segments that are economically viable in case of applying the slow formation and dissolution strategy.

The results indicate a significant reduction of suitable platooning segments when taking into account feasibility and even more when taking into account economic viability. With respect to economic viability, the medium formation/dissolution strategy results in higher numbers of platooning segments than the slow and fast strategy (medium > slow > fast).

Findings within the data analysis for the route from Pasching to Guntramsdorf are:

- Feasibility assessment
  - Some road segments are not eligible due to falling below the required length for platoon formation and dissolution.
  - The prohibition of platooning near on-/off-ramps significantly affects the delta between "Platooning allowed" and "Platooning feasible", e.g. Fig. 12.1 delta for number of road segments in scenario A = 18 VS delta in scenario F = 2.

- Economic viability assessment

Route analysis "Pasching -> Guntramsdorf"
with respect to platooning feasibility and economic viability

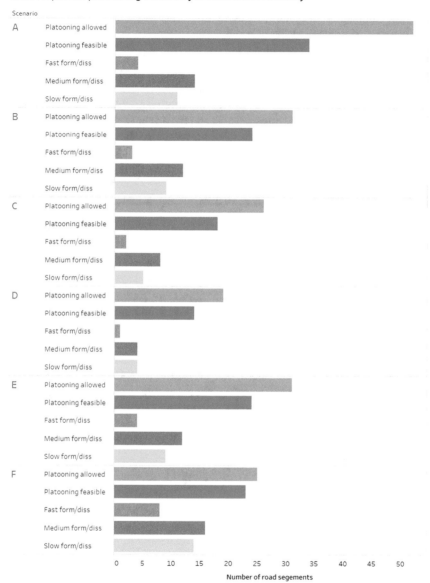

Fig. 12.1 Fuel efficiency assessment—example

– The economic viability assessment additionally decreases the number of road segments suitable for truck platooning.
– Medium platoon formation and dissolution strategies provide the highest number of economically viable platooning road segments.

Subsequent to the assessment of feasibility and economic viability, an analysis of potential costs and emission savings was conducted. In order to assess the fuel saving potential for the given routes, the fuel saving model presented in Fig. 5.10 was applied. The maximum average fuel savings are depicted in Fig. 12.2 with respect to the direction, the regulation scenario (A-F) and the driving behaviour. This analysis confirmed that in most instances the medium formation/dissolution strategy for a 3-truck platoon driving at an intra-platoon distance between 1 and 1.5 s at a speed of 80 km/h is suitable. The maximum achievable fuel saving is 4.83% for all three trucks in a platoon driving from Pasching to Guntramsdorf in scenario F applying the slow formation/dissolution strategy at an intra-platoon distance of 0.5 s and speed of 80 km/h.

In addition to the maximum fuel savings, configurations that lead to minimum fuel savings were analysed. As depicted in Fig. 12.3, the fast formation/dissolution strategy at an intra-platoon distance of 1 s leads to minimum average fuel savings in a platoon of three trucks. The minimum saving for the route Pasching to Guntramsdorf in scenario F applying the fast formation/dissolution strategy at an intra-platoon distance of 1 s leads to fuel savings of 2.53%. This minimum saving value is 2.3% less than the maximum achievable value of 4.83% given above.

The analysis of the fuel reduction also provides a basis for assessing potential savings in CO_2 emissions. Based on the consumption values and a factor for translating fuel consumption in emissions, one may illustrate potential CO_2 savings. In Fig. 12.4, maximum potential CO_2 savings for a three-truck platoon are depicted for the different regulation scenarios and transport routes. The emissions are calculated based on the emission factor provided by [1].

The depicted values may provide a basis for fleet operators to project the fuel and emission savings for their fleet operations. Doing so, they may assess how truck platooning can contribute to sustainable transport operations. Furthermore, the saving potentials might be used when authoring emission reports related to their logistics operations.

12.2 Traffic Efficiency Assessment

In this section, the effects of energy-efficient automated convoys on traffic efficiency and traffic control are analysed and measures are derived. Platooning allows groups of trucks to safely drive at small headways, thereby reducing their overall length as a unit. Depending on the exact platoon specifications (availability of C-ITS-equipped trucks, percentage of trucks on the road, ability to form inter-company platoons, exact platoon headway, etc.), the throughput of existing infrastructure can be increased.

Max % of (average) fuel savings in a 3-truck platoon

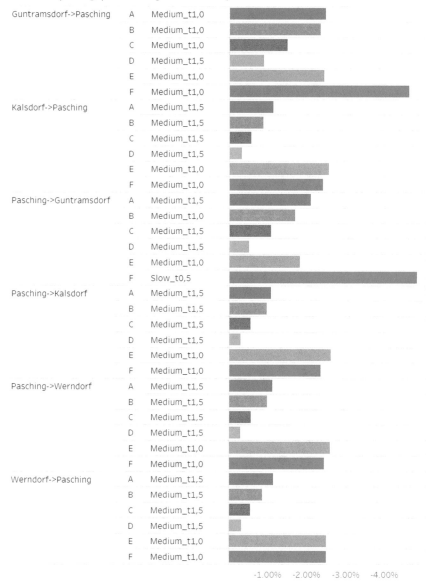

Fig. 12.2 Maximum average fuel savings in 3-truck platoon—fleet operator example

Min % of (average) fuel savings in a 3-truck platoon

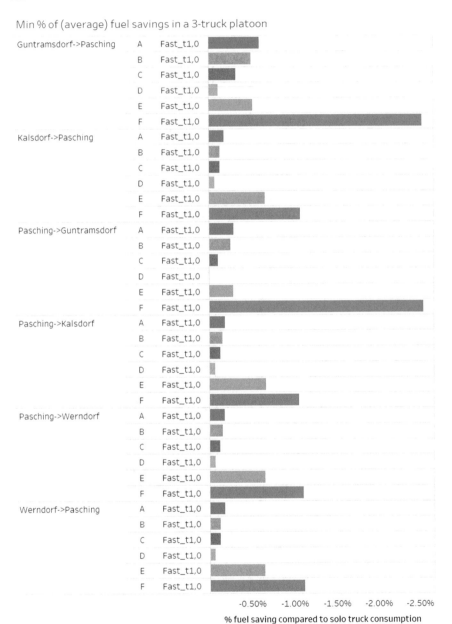

% fuel saving compared to solo truck consumption

Fig. 12.3 Minimum average fuel savings in 3-truck platoon—fleet operator example

Overall emission savings for a 3-truck platoon

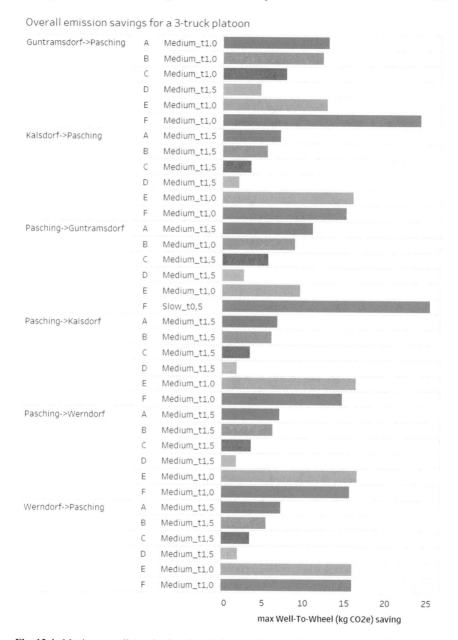

Fig. 12.4 Maximum well-to-wheel savings in 3-truck platoon—fleet operator example

However, the number of unknown parameters for platooning specifications requires a thorough analysis based on Latin hypercube sampling. The resulting multidimensional data of achievable traffic throughput may then be used for detailed analyses. At the point of writing, many truck platooning parameters are completely unknown and will only become available step by step in future. However, the sampling results are complete in the sense that a result for each conceivable (at the point of writing) combination of parameters is available and can therefore readily be evaluated in detail.

The method to investigate a potential increase in traffic volume has been introduced in Sect. 9.2. The relationship between the input parameters and the attainable increase in traffic volume are presented. Following, the evaluation and interpretation of results are provided.

Two general approaches that are represented by different sets of assumptions are analysed. Therefore, a set of generic assumptions ("mixed platooning") is proposed, in order to establish the true theoretical upper bound of the influence of platooning, C-ITS measures and their expected impact. Generic assumptions are:

- Any vehicle can be a part of a platoon.
- Platoons require back-to-back driving of vehicles with C-ITS capabilities.
- A platoon can contain an arbitrarily large number of vehicles.

This set of assumptions will not be met in the near foreseeable future. Nevertheless, it is vital to quantify a theoretical upper limit of improvement in order to establish a reasonable roadmap for future research and implementation efforts. In case that the theoretical upper limit represents only an insignificant improvement, one can decide at this point that further research in this area is futile and resources are better spent at issues that provide a significant impact. The second set of assumptions ("truck platooning") is:

- Only trucks can be a part of a platoon.
- Platoons require back-to-back driving of trucks with C-ITS capabilities.
- A platoon can contain a maximum of three vehicles.

In addition to the parameters mentioned, the vehicle lengths and vehicle type distributions influence traffic efficiency significantly. For this reason, current highway control measurement data provides information on the state of vehicle distributions and lengths on Austrian highways. In order to gauge theoretical limits, the range of these parameters is extended well beyond currently measured values. However, the extension considers physically possible limits, so as to account for possible future development. The described parameters are summarised in Table 12.1.

In the following, traffic efficiency assessment cases are detailed. Note that C-ITS capability is a stochastic variable. The state of the art is represented by conventional trucks that have *no C-ITS capability*. In future, a growing fraction of trucks will be sufficiently equipped, and finally, in the far future, all trucks may be equipped with C-ITS support. Therefore, the range of this particular parameter is between zero and one. At any given time in future, e.g. when a survey shows that a certain fraction of

Table 12.1 Parameter for traffic efficiency assessment

	Distr. of vehicle lengths (m)	Distr. of vehicle types	Platoon type
Case 1	From measurement data	From measurement data	Truck platoons only
Case 2	Conservative assumption $L_{LKW} = 18.75\,m$, $L_{PKS} = 4.5\,m$	Stochastic sampling	Truck platoons only
Case 3	From measurement data	From measurement data	Mixed platoons
Case 4	Conservative assumption $L_{LKW} = 18.75\,m$, $L_{PKS} = 4.5\,m$	Stochastic sampling	Mixed platoons

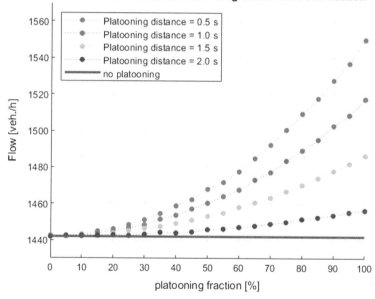

Mean flow for measured vehicle length dist. and truck fraction

Legend:
- Platooning distance = 0.5 s
- Platooning distance = 1.0 s
- Platooning distance = 1.5 s
- Platooning distance = 2.0 s
- no platooning

y-axis: Flow [veh./h]
x-axis: platooning fraction [%]

Fig. 12.5 Traffic volume increase for true vehicle length and type distributions and different intra-platoon headways (case 1)

trucks is equipped with C-ITS capabilities, the results of this particular investigation can be applied with the given value for that parameter.

For truck platooning, based on actually measured vehicle lengths and type distributions, the increase in traffic volume is shown in Fig. 12.5. The measured truck fraction is approximately 11%. For fractions of C-ITS capable trucks (represented by platooning fraction) < 50%, the increase of traffic volume is < 1%. The maximum achievable increase in traffic volume of about 7.5% is reached at 100% C-ITS capability and the smallest possible time headway of 0.5 seconds. This is in itself not

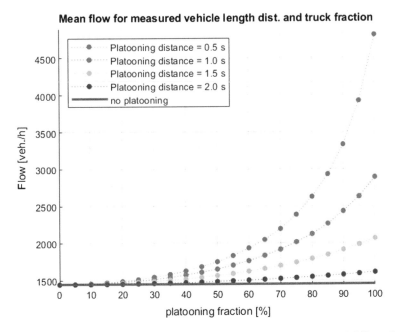

Fig. 12.6 Traffic volume increase for true vehicle length and type distributions and different intra-platoon headways (case 3)

a very impressive number, given that the imaginary setting is placed well in future. However, it also indicated that the notion of pure truck platoons may be limiting.

For the "mixed platooning" case, traffic volume increase is significantly higher. In Fig. 12.6, the same plot is shown for mixed platooning, representing case 3 sketched in Table 12.1.

Similar result plots are useful for the cases 2 and 4, where the vehicle distribution is sampled and a Monte Carlo approach is utilised to generate results.

For truck platooning in case 2, a particular situation for the truck fraction of 25% and a platooning fraction of 80% is shown in Fig. 12.7. An increase of traffic flow of about 11% can be achieved for a small time headway of 0.5 s and the assumptions stated above.

The results are more promising for mixed platooning as indicated in a similar plot in Fig. 12.8.

Summarising, the achievable increase in traffic volume, i.e. the potential to increase traffic efficiency, depends significantly on the type of platoons that are considered (mixed platoons or truck platoons only), and the fraction of vehicles that have the technical capability to participate in platooning. For all practical purposes in the near future, the fraction of C-ITS capable vehicles in general and trucks in particular may not be large enough to increase efficiency noticeably. However, traffic efficiency is not the only target for platooning as described in detail in Sect. 4.1. The results presented provide an overview of what can be achieved under particular circumstances represented by a set of assumptions and the corresponding range of

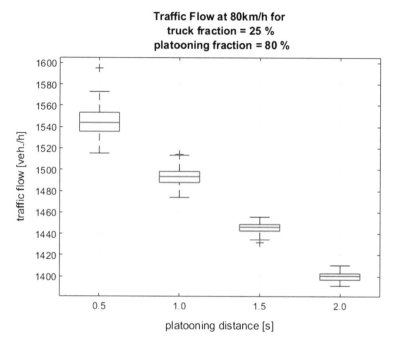

Fig. 12.7 Traffic volume increase for constant vehicle length, stochastic type distribution and different intra-platoon headways for truck platooning only (case 2)

parameters. Even if traffic efficiency by itself is not the driving force for further truck platooning research and implementation, a positive impact can be expected.

12.3   C-ITS Assessment for Dynamic Traffic Control

C-ITS enables dynamic truck platooning control on highways and supports further increasing efficiency. Instead of statically defining restrictions for truck platooning on highways, dynamic information regarding current restrictions for a highway segment might be provided via C-ITS. The restrictions may depend on factors such as traffic density, hazardous events, day time or weather conditions. Applying a dynamic C-ITS-based approach for platoon management on highways allows to combine truck platooning regulation scenarios (cf. Fig. 3.3). Such a combination supports increasing the number of "platooning-eligible" highway kilometres as depicted in Fig. 12.9. The figure illustrates the following situation:

- Route: Austrian highway A1, A21, A2 from Linz to Vienna.

  – Highway on-ramp Ansfelden to exit Guntramsdorf.

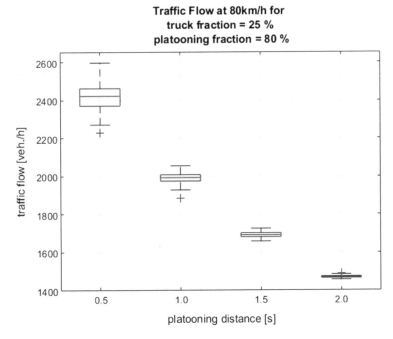

Fig. 12.8 Traffic volume increase for constant vehicle length, stochastic type distribution and different intra-platoon headways for mixed platooning (case 4)

- Platooning restrictions:

  - Near cities, i.e. Linz, St. Pölten, Vienna, the scenario A "Still safe" rules.
  - In between cities scenario F "Allow ramps" rules.

- Information with respect to the ruling scenario is communicated to vehicles via C-ITS.

Combining both, scenarios A and F, as depicted in Fig. 12.9 allows to increase the number of platooning-eligible kilometres from 116km (exclusively A) to 139km (combination A+F in the given situation). In other words, if on-/off-ramps are allowed on highway segments in between cities from Linz to Vienna, the number of platooning-eligible kilometres may be increased by nearly 20%. As stated above, the dynamic application of different platooning regulations should take into account influencing factors such as the highway topology and the real-time traffic situation.

Another application case could consider dynamic management of tunnel restrictions. In Austria, the A9 from Upper Austria to Styria comprises a high number of tunnels. Therefore, potential savings could be gained in case platooning is considered safe in the given tunnels in certain traffic situations. Figure 12.10 illustrates the following situation:

- Route: Austrian highway A1/A9 from Linz to Graz

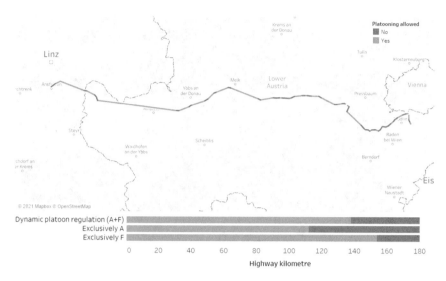

Fig. 12.9 Dynamic C-ITS-based platoon management—example 1, base map and map data from OpenStreetMap, © OpenStreetMap contributors under the CC-BY-SA license, https://www.openstreetmap.org/copyright

- Highway on-ramp Ansfelden to exit Werndorf.

- Platooning restrictions:

  - In Upper Austria scenario B "Astra Study -no Bridges" rules.
  - In Styria E "Astra Study -no Bridges -no Tunnels" rules.

- Information with respect to the ruling scenario is communicated to vehicles via C-ITS.

Described parameters are summarised.

Combining both, scenarios B and E, as depicted in Fig. 12.10 allows to increase the number of platooning-eligible kilometres from 54 km (exclusively B) to 92 km (combination B+E in the given situation). In other words, if tunnels are allowed on highway segments in Styria on the route from Linz to Graz, the number of platooning-eligible kilometres may be increased by 70% compared to the "exclusively B scenario". The highest number of platooning-eligible kilometres could be gained in the "exclusively E scenario". If this scenario would rule in Upper Austria and Styria, the number of platooning-eligible kilometres would be 120km, which is more than half of the overall route.

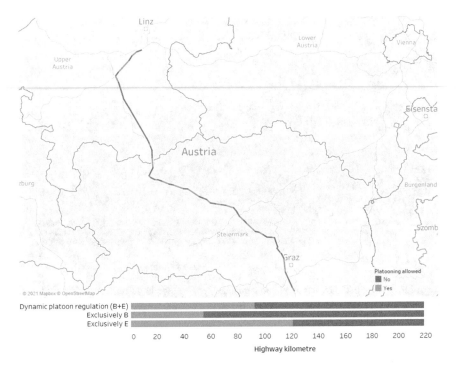

Fig. 12.10 Dynamic C-ITS-based platoon management—example 2, base map and map data from OpenStreetMap, © OpenStreetMap contributors under the CC-BY-SA license, https://www.openstreetmap.org/copyright

12.4 Conclusion

Efficiency represents a crucial element of truck platooning. Efficiency may relate to actual fuel savings in a platoon as well as to traffic efficiency and control. Fuel savings depend on diverse aspects such as the following distances of trucks in platoon, or the additional fuel consumption for platoon formation and dissolution. In this chapter, a typical route of an Austrian fleet operator was assessed with respect to the feasibility of platooning and the economic viability of platooning segments. The results with respect to feasible routes indicate that depending on the scenario the feasibility of allowed platooning road segments varies; e.g. for the the transport route "Pasching–Guntramsdorf", the share of feasible segments ranges from 92% (scenario F) to 65% (scenario A). The highest share of economic viable road segments is gained when applying a medium formation and dissolution strategy for truck platoons. Restricting truck platoons on on-/off-ramps within the given assessment situation decreases the number of road segments allowed for platooning and leads to a higher number of shorter road segments allowed for platooning that are economically not viable. Therefore, when restricting truck platooning one should also consider the resulting fragmentation of platooning allowed/not allowed road segments. A high number of short segments is considered less efficient compared to a lower number of long platooning segments. Furthermore, applying dynamic C-ITS-based traffic control approaches may increase to number of platooning kilometres compared to static traffic regulations.

Aside from the fuel efficiency assessment, a traffic efficiency and control assessment was conducted in this chapter. The results indicate that the achievable increase in traffic volume depends significantly on the type of platoons that are considered (mixed or truck platoons only), and the amount of vehicles that have the technical capability to participate in platooning. For all practical purposes in the near future, the amount of C-ITS capable vehicles in general and trucks in particular may not be large enough to increase traffic efficiency noticeably.

Reference

1. Greene S, Lewis A (2019) Global logistics emissions council framework for logistics emissions accounting and reporting, version 2.0

Elvira Thonhofer has received her Masters Degree in Mechanical Engineering from theVien-naUniversity of Technology. Her research interests include traffic modelling, simulation and control, intelligent transportation systems and automation. Elvira has worked on national research projects and her academic contributions are published in relevant journals and conferences.

Matthias Neubauer is a professor at the University of Applied Sciences Upper Austria for logistics information systems. His research interests cover human-computer interaction, intelligent transportation systems as well as cooperative, connected and automated mobility. Matthias is involved in international and national research projects and teaches in the master's program digital transport and logistics management classes such as process management, distributed logistics systems or geoinformation systems.

Florian Hofbauer is a research associate in the research group 'hyperconnected logistics systems' at the University of Applied Sciences Upper Austria. His research focus is on sustainability assessment of intelligent transport systems, automated mobility, alternative fuels and sustainable transport systems.

Part III
Towards Cooperative Truck Platooning Deployment

Part III outlines paths towards cooperative truck platooning deployment. In doing so, traffic safety-related issues in general as well as specific issues in Austria are discussed as a main prerequisite for automated driving. Moreover, taking a road operator's view, business model aspects relevant for the C-ITS deployment are assessed to reveal sustainable next deployment steps. The remainder of this part investigates advanced powertrain systems to increase sustainability of truck platoons and impacts of truck platooning research on the European innovation system. Part III concludes with a general discussion on lessons learnt and open questions regarding CCAM.

Chapter 13
Road Safety Issues Related to Truck Platooning Deployment

Susanne Kaiser, Martin Winkelbauer, Erwin Wannenmacher, Philipp Blass, and Hatun Atasayar

Abstract The benefits of platooning to road safety are oftentimes inferred based on the assumption of positive effects attributed to advanced driver assistance systems (ADAS). However, the potential to significantly reduce car crashes is just one of many aspects to be considered. The Connecting Austria project was committed to examining road safety issues from various perspectives within Austria. The legislative situation in Austria regarding public tests of automated driving systems was reviewed and requirements discussed. Furthermore, an assessment of the readiness of 700 km road infrastructure was assessed by means of an adapted Road Safety Inspection and recent heavy goods vehicle (HGV) accident figures on motor and expressways were discussed. Eventually, the distance at which HGV should operate cooperatively is a road safety issue demanding for consideration of other road users. Car drivers' subjective tolerance of gap sizes between trucks can serve as an important indicator to answer this question. An on-road study aiming at operationalising the individual gap acceptance is outlined.

Keywords Road safety · Platooning · Gap acceptance · ADAS · Road safety inspection

13.1 Introduction

While decreasing transportation costs and reducing greenhouse gas emissions are explicit goals of platooning, many truck platooning initiatives so far have neglected the potential effects on road safety or merely assumed positive effects without providing sound evidence. Therefore, *Connecting Austria* committed to focusing on potential road safety costs and benefits from different perspectives. Four distinct topics were subject to the analyses, considering legal, statistical, infrastructural and human aspects. (1) The legal situation in Austria regarding automated driving in general and

S. Kaiser (✉) · M. Winkelbauer · E. Wannenmacher · P. Blass · H. Atasayar
Austrian Road Safety Board (KFV), Vienna, Austria
e-mail: susanne.kaiser@kfv.at

© The Author(s) 2022
A. Schirrer et al. (eds.), *Energy-Efficient and Semi-automated Truck Platooning*,
Lecture Notes in Intelligent Transportation and Infrastructure,
https://doi.org/10.1007/978-3-030-88682-0_13

platooning in particular was reviewed, and requirements for testing platooning as a use case are discussed. (2) HGV accident figures of recent years are used to explore the safety potential of platooning: in the period of 2014–2018, inattention/distraction and insufficient safety margins were the leading causes of crashes, comprising two-thirds of accidents involving HGVs on Austrian motorways and expressways. The promise of these figures, indicating great potential to prevent injuries and damages, is tied to the assumption that automated and cooperative platooning systems are widespread and active at all times and failsafe. Furthermore, when accounting for safety gained through ADAS, the benefits of platooning itself are less obvious and harder to quantify. It will become increasingly important to not only document actual crashes but other relevant indicators for safe platooning such as the interaction between human and machine too. (3) Road infrastructure conditions on relevant Austrian motorways were assessed in terms of suitability for platooning by means of adapted Road Safety Inspection (RSI). One part of a traditional RSI is an on-road assessment. To this effect, more than 700 km on motorways, expressways and federal roads were evaluated, with a stretch of 190 km on the motorways A1 and A21, being used as an example for detailed discussion in this chapter. (4) Eventually, an outline of an on-road study with 96 participants is given, aiming at explaining car drivers' subjective acceptance of gaps when merging between two HGV, which is an important measure for determining the minimum and maximum distance at which trucks should operate cooperatively.

13.2 Legal Aspects for Platooning in Austria

For the assessment of legal requirements for platooning in Austria, a platoon of two to three HGVs, equipped with advanced driver assistant and control systems as well as car-to-car communication technically allowing automated close following of the second and third vehicle on motorways, is assumed. Furthermore, it is assumed that each of the following vehicles is manned with conductors who supervise the vehicle and traffic and who take over manual driving if needed.

13.2.1 Acquiring a Test Permission According to the Austrian Regulation on Automated Driving

Testing (partially) automated systems on road is a necessary step towards eventual admission and registration in Austria or may serve mere research purposes. Prior to testing on public roads, virtual or on-road tests on private property roads are required. Testing on public roads can also be staged, for example, with tests under ideal conditions (good weather and road conditions) followed by increasingly adverse conditions.

Legal requirements for automated driving in Austria are regulated in the *Automatisiertes Fahren Verordnung—AutomatFahrV* [1]. While the *AutomatFahrV* regulates the requirements for automated systems to be tested on public roads, § 102 *KFG* (Austrian act on motor vehicles) provides the possibility for absolving the driver for certain responsibilities and transfers them to assistance and automated systems (e.g. keeping at least one hand on the steering wheel) per ordinance, in the first place.

The Austrian testing procedure of (partially or highly) automated vehicles is overseen by the Department for Transport (Federal Ministry "Climate Action, Environment, Energy, Mobility, Innovation and Technology"). A fundamental prerequisite is that the driver remains ready to resume control at all times—in accordance with the *Vienna Convention on Road Traffic*. Moreover, the automated systems must have the ability to comply with specifications of the *StVO* [2] (Austrian road traffic act), the *EisbKrV* [3] (Austrian regulation for level crossings) and the *IG-L* (Austrian act on air pollution control). Vehicles must be roadworthy and safe to operate and must meet certain other criteria for admission, such as insurance and data recording requirements. The *AutomatFahrV* contains provisions for (1) driver assistance systems which are already approved and in series production (parking assistant, motorway assistant with automatic lane guidance) and (2) systems which are used for test purposes and which can be assigned to one of three predefined use cases (autonomous minibus, motorway pilot with automatic lane changing, autonomous military vehicle and motorway assistant with automatic lane changing). The Federal Ministry "Climate Action, Environment, Energy, Mobility, Innovation and Technology" is the authority which issues the permission to test on public roads for a limited period of time, given the requirements are met. By the end of the testing period, a full report on the knowledge gained has to be submitted to the ministry and immediate notice has to be given in cases of critical situations or crashes including the underlying cause. Furthermore, the Department for Transport issued a "Code of Practice" [4] which aims at additionally increasing the safety level of testing.

13.2.2   Does the Current Law Facilitate Testing of Platoons on Austrian Roads?

One of the test use cases according to the *AutomatFahrV* is "motorway assistant with automatic lane changing", which refers to systems which can guide the longitudinal movements (accelerating, decelerating, complete stop, keeping distance) and lateral movement (lane keeping, lane changing, overtaking) of the vehicle. The respective system can only be tested by vehicle manufacturers, system developers and research institutions on motorways and express roads (once the vehicle has merged into the flow of traffic) in a defined set of vehicle types (M1, M2, M3, N1, N2 and N3 Motor Vehicle Act). Furthermore, the system must have been tested for at least 10,000 km on private property and have proper insurance through a liability insurer. A trained test driver is indispensable and must be able to resume the driving tasks and stop the automated system at all times. In case of an emergency or critical situation, the obligatory emergency mechanism also is activated, and the driver resumes control. They have to resume control well ahead of an exit as well. A black box has to be installed. The obligatory self-commitment to the "Code of Practice" aims at risk mitigation (contingency plan, risk management, etc.). Handling a critical situation must be subject to the training of test drivers, who are obliged to hold a valid driving licence and have several years of experience driving the respective vehicle.

This use case may apply to the system which was subject to the Connecting Austria project (HGV). The specifications above are indicators that the project's endeavour is potentially in line with the motorway assistant use case. However, the ordinance refers to systems which feature longitudinal and lateral control, while the Connecting Austria project aims at building, maintaining and resolving platoons through intelligent V2V without targeting automated lane change and overtaking manoeuvres. The vehicles are supposed to drive back to back with reduced distance between them. Carefully considering the aim and focus of platooning and the legal texts lead to the conclusion that platooning and the motorway assistant are two different use cases with different aims, especially since (1) platooning involves more than one vehicle, (2) the requirements for road safety and infrastructure are different (short distance between trucks), (3) the platooning system aims at additional tasks (V2V) and (4) the road types for platooning ideally are not limited to motorways, to name but a few.

13.2.3 Requirements for Platooning Tests in Austria from a Legal Point of View

If a testing scheme does not fit the current regulations, potential test cases can be suggested to the *Kontaktstelle Automatisierte Mobilität* (contact point for automated mobility). Should the proposal be sufficiently legally and technically sound, the ministry can issue an amendment to the ordinance, accordingly.

However, one major barrier on the path towards testing of truck platooning relates to the *StVO* (Austrian Road Traffic Act). The *AutomatFahrV* requires conformity with the *StVO* which regulates under § 18 Abs 1 *StVO* a safe distance that allows a following vehicle to stop safely in case of sudden braking of the lead vehicle. Thus, a safe distance is dependent on speed, road and vehicle conditions, driver attention as well as the current distance. Generally, a two-second rule applies: two times the reaction distance, which increases especially for speeds over 100 km/h (three seconds) and hazardous circumstances (three to four seconds time headway) (e.g. [5]). Moreover, vehicles of bigger dimensions—such as trucks—are required to keep a minimum safety gap of 50 m to vehicles of the same kind on rural roads and motorways (§ 18 Abs 4 *StVO*).

Since platooning per definition is expedient when the distance between the single trucks is reduced, conformity with the *StVO* cannot be assumed. In order to resolve this discrepancy, an amendment of the *StVO* (§ 18 Abs 4) towards an exemption for

Table 13.1 Causes of HGV accidents on Austrian motor and expressways from 2014 to 2018 (top eight = 96% of accidents)

Cause of HGV accidents	Total (%)
Inattention and distraction	51
Insufficient distance to vehicle in front	12
Inadequate speed	9
Fatigue and sleepiness	8
Right of way violation	5
Technical defect and cargo security	4
Overtaking manoeuvre	4
Substance impairment	3

Data source Statistik Austria—data.statistik.gv.at [6] under the CC-BY-4.0 license

truck platooning would be necessary. Furthermore, from a road safety perspective it remains to be considered if novel traffic signs should be incorporated into the *StVO* and if vehicles should be prohibited from merging between the vehicles in a platoon.

13.3   Considerations for the Safety Potential of Platooning

One of the major advantages (or hopes) of driving assistant systems such as electronic distance control is the lack of some of the human drivers' flaws and weaknesses like error-proneness due to fatigue, sleepiness, distraction, substance impairment, etc. Electronic control will also be much more accurate with much less variation compared to human driving. For the purpose of estimating the crash reduction potential, HGV accidents of the past years (2014–2018) on Austrian motor and expressways will be looked at, considering vehicles >3.5 tonnes [6].

Usually, the accident type is one of the standard parameters to look at. However, due to the focus on motor and expressways, it is not surprising that the large majority of 82% of HGV accidents is classified as same direction accidents involving other road users and another 9% are single vehicle accidents. Looking at the causes of HGV accidents is more revealing (Table 13.1). The top five accident causes are linked to human errors, with inattention and distraction being the lead causes, making up around half of HGV crashes. Ranked number six, technical defects or insufficient cargo security represents only 4% of the analysed accidents.

Although it should be noted that the classification of the accident cause is based on the law enforcement agent's assessment and not a thorough accident investigation, the potential of crash reduction when eradicating human errors seems to be vast. In the context of automated and connected driving in general, it will become more and more important to adapt the accident classification scheme accordingly. However, when anticipating a potential admission of platooning, a more detailed reporting of incidents regarding the systems, the driver and the interaction with the environment is desirable. Asare et al. for example proposed in [7] twelve safety performance measures for truck platooning, including not only rates of crashes, near-crashes and safety critical conflicts but also the number and types of system failures, the conditions under which they occur, fails to notify the drivers about loss of control, V2V signal loss, disengagement of the driver, episodes and levels of fatigue, levels of vigilance and distraction of the driver, cut-ins of other vehicles.

Overall, the potentials for road safety due to platooning stand and fall with the reliability of the implemented ADAS and the proper use by operators. Based on these assumptions, platooning could substantially contribute to safer HGV traffic.

13.3.1 Safety Potential of Platooning Compared to Existing Safety Assistance Systems

Previous observations have clearly shown that there is considerable potential in the use of platooning technology regarding the avoidance and mitigation of traffic accidents. However, the extent of the improvement depends strongly on the reference scenario. To answer the question of a potential safety gain through platooning, the underlying assumptions must be clearly defined. For this discussion, a platoon is assumed consisting of a lead vehicle and two following vehicles, all equipped with the same accident preventing assistance systems as three HGV driving alone. Differences are therefore only taken into account to the extent that they result from the "platooning" function, i.e. the electronic connection and the possibility for the drivers in the following vehicles to divert their attention away from overseeing the traffic environment. When controlling for the effect of ADAS used in platoons, the safety benefits of platooning lie in shared information between the trucks. The following vehicles in the platoon have a head start concerning relevant information: they "know" about dangers at almost the same time as the lead vehicle and therefore react earlier and, if necessary, even use this time advantage flexibly for themselves or other road users following the platoon (V2V given). The reduced collision speeds due to faster reaction in case of rear-end collisions would result in decreased severities of collisions.

13.4 Assessment of Road Infrastructure with Respect to Safe Platooning

Providing safe road infrastructure is an important goal of public agencies. Road Safety Inspection (RSI) is one of many tools used to identify potential deficiencies and safety risks in the road network and to address them with appropriate countermeasures to prevent accidents or to mitigate accident consequences [8]. In 2011, the EU Directive 2008/96/EC (*Infrastructure Directive*) was implemented in Austria by

amending *BStG* (Austrian Highway Act), which now details the RSI and requirements for becoming a certified inspector (RVS 02.02.35 Certification of road safety
inspectors). The procedure of the RSI is four-part: (1) "preparatory work such as
review of the existing documents, collection of accident data, etc.", (2) "site visit
including discussions with people responsible for the road", (3) "creation of the RSI
report" and (4) "implementation of the proposed measures, monitoring" [8, p. 4].
Checklists are available for different road types to support identifying safety-relevant
features.

For the purpose of assessing Austrian roads with regard to the suitability for
platooning, a team of experts of the Austrian Road Safety Board (KFV) adapted the
RSI to the project's needs. In a first step, the expert team defined a set of criteria which
can serve as a basis for future evaluation of road sections and which are required for
safe platooning in the experts' view:

- Minimum of two lanes (two lanes are considered sufficient if all of the other criteria
  apply).
- Shoulder width of ≥3 m.
- Dissolution of platoons 1000 m ahead of tunnels.
- Dissolution of platoons 1600 m ahead of construction sites.
- Dissolution of platoons 1000 m ahead of hazard points such as accident black
  spots.
- Route is characterised by only few bends (curve radios >1000 m).
- Dissolution of platoons 1000 m ahead of motorway junctions (on-/off-ramps),
  motorway stations and rest areas (the merging of traffic requires complex orien-
  tation tasks and masking information aiming at support of orientation for drivers
  should certainly be avoided).
- Sufficient length of acceleration and deceleration lanes.

The following aspects of a classic RSI were assessable within the project: (1)
presence and visibility of lane markings and influence of moisture, fog, wind, etc.,
(2) distance between direction signs and motorway exit (distances potentially have
to be increased), (3) distance between motorway junctions, (4) road's gradient and
(5) accident analysis including accident black spots (beforehand assessment).

13.4.1 Performance of the On-Road Assessment

Subject to the assessment were sections of the Austrian motorways A1, A2, A7, A10,
A21 and A25 as well as selected expressways and national roads which represent
some of the typical routes of the carrier and project partner Transdanubia. For the
analysis in this contribution, the following route is included (one direction only, from
East to West): A21 motorway junction Vösendorf till junction Steinhäusl, continuing
on A1 till junction Haid.

The respective road network was segmented into meaningful geographic units
marked with motorway kilometrage before the on-road assessment started. Both

Table 13.2 Characteristics of the assessed segmented route on the Austrian motorways A1 and A21

	Suitable for platooning		Total
	Yes	No	
Total length (km)	82.74	107.8	190.54
Number of segments	32	36	68
Range of segment length (km)	0.3–10.2	0.2–4.7	0.2–10.2
Mean of segment length (km)	3.37	2.3	2.8
Median of segment length (km)	2.05	1.9	1.95

directions were considered and part of the on-road assessment. The evaluation was based on the on-location sighting as well as video recordings and was conducted in December 2018. It is assumed that a potential platoon drives on the rightmost lane. As detailed in the list of criteria, above sections in the proximity to motorway junctions (on-/off-ramps) were excluded from a positive evaluation at the outset.

13.4.2 Analysis of Road Segments and Considerations for Platooning

A total stretch of 190.54 km was assessed in terms of suitability for platooning, segmented into 68 units. Table 13.2 gives an overview of the main characteristics of the route, for both positively and negatively evaluated route segments. About half of the segments are suitable for platooning; this half accounts for 43.6% of the route in length. Suitable stretches show a mean of 3.37 km and a median of 2.05, with a range of 0.3–10.2 km. See Fig. 13.1 for the distribution of segment lengths.

Since the segments were chosen according to motorway junctions as points of reference, there are no segments back to back which are suitable for platooning: each positively evaluated segment is interrupted by either a motorway junction or service stations/rest areas. The longest identified segment suitable for platooning on the reference route therefore is 10.2 km long and is located on the motorway A1 between the service stations St. Pölten and the junction Loosdorf. Not accounted for was the distance needed for building and dissolving a platoon. Dependent on the initial speeds of the HGVs as well as the compression and decompression pace and protocol, the processes combined require distances between 0.8 and 3.5 km (see Tables 11.3 and 11.4). Nine out of 32 route segments are 5 km or longer, and 20 are 3.6 km or shorter. All the route segments which resulted in a negative evaluation are characterised by on-/off-ramps (motorway junction, service station, rest area or a combination of those three). Furthermore, for three of those road segments a spike in truck accidents between 2015 and 2017 was observed (junction St. Pölten, Linz/Ansfelden and Traun/Haid).

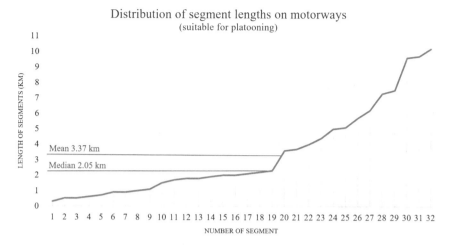

Fig. 13.1 Distribution of segment lengths for positively evaluated route segments on Austrian motorways A1 and A21 (between A21 junction Vösendorf and A1 junction Haid), $n = 68$ motorway segments

13.5 Gap Acceptance of Car Drivers for Merging Between Trucks

One of the questions within the platooning discourse is the ideal distance between the trucks of a platoon. Alongside economic and environmental considerations, this is also an important road safety question. Not only in terms of potential malfunction of the used technical equipment but also for the perception of other road users. Car drivers for example have to decide at which gap sizes between trucks they would merge onto the rightmost lane for the purpose of exiting the motorway or merge into traffic coming from an on-ramp. While it is argued above that a platoon should dissolve well ahead of a motorway junction, it also is undesirable to have car drivers merge between the trucks and thereby break off the connected platoon. While it is known that some drivers are more and some drivers are less inclined to take risks, the roads must be safe for all drivers. To gain knowledge about the drivers' decision-making regarding the acceptance of different gap sizes between trucks, an on-road driving study was conducted in summer 2020. However, in this contribution only a basic outline of the study design will be given since the data analyses were still ongoing.

About one hundred participants were recruited to drive an instrumented vehicle on an approximately one-hour route in and around Vienna with stretches characterised by a high share in trucks due to a nearby goods distribution centre. Besides recording a multitude of CAN bus-derived driving parameters such as speed, acceleration and deceleration, the test vehicle was equipped with two roof-mounted lidar sensors for measuring gap sizes between trucks. Before the test drives started, participants were informed of their task during the test drive, which was to assess whether they hypothetically would merge from the second lane between two trucks upon request of the test administrator on the passenger seat (leaving the participants the options of a definitive "yes" or "no" and the third category "in case of need"). The tasks of the test administrator were to give directions to the participants, to identify gaps of interest between trucks on the expressway as well as to record the participants' answers and the timestamps for their decisions. Data of $n = 96$ participants was successfully collected and is subject to comprehensive analyses, considering a wide range of covariates such as speeds, sociodemographic factors, private car use, risk-taking proneness, comfort while driving, truck types and others.

13.6 Discussion

The discussed topics in this contribution are indicative of the fact that road safety is a multi-layered concept requiring interdisciplinarity: legal, technical, infrastructural and human aspects were touched upon, which represent only a few factors to be considered for the safe introduction of platooning. With this in mind, this text does not claim to provide an exhaustive overview but rather a discussion of selected road safety topics which were subject to the *Connecting Austria* project.

On the way to testing platooning in Austria, the required safety margin of at least 50 m between trucks seemingly represents one of the largest obstacles at this point from a legal perspective. This required distance is defined in the *StVO* (Austrian road traffic act), which cannot easily be circumvented by the Austrian Ordinance on Automated Driving. However, there are promising and realistic approaches to overcoming this obstacle. A more challenging discussion lies ahead concerning a potential market introduction, with potential impact on a variety of federal and EU laws and regulations, such as the *KFG* (Austrian Act on Motor Vehicles), *FSG* (Austrian Act on Driver Licensing), the *Vienna Convention on Road Signs and Signals*, regulations on driving time and resting periods, etc. Defining requirements and behavioural guidelines for drivers as well as parameters for the road infrastructure will be especially interesting. The latter will be crucial not only for road safety but also for the profitability of platooning in Austria, since long road stretches between single motorway junctions are rather rare. Allowing other vehicles to safely enter and exit motor or expressways is certainly a priority. To which extent this can be supported by means of V2V or I2V as a mitigation measure needs to be investigated further.

Although it is fair to say that road safety is not the primary motivation for proponents of platooning, it is worth discussing potential safety benefits. Human errors are without a doubt a leading cause of road accidents, with inattention and distraction alone accounting for 51% of HGV crashes on Austrian motor and expressways between 2014 and 2018. Thus, it can be assumed that platooning has the potential to prevent accidents. However, one unknown factor in the equation is the reliability of systems that control the connected/automated driving function as well as the assumed penetration rate. Furthermore, the effect most likely would be attributable to ADAS, meaning that single trucks with the same technical equipment would perform just as well. A safety consideration unique to platooning is the shared information within the platoon, which is not dependent on the platoon being connected to outside information systems for a head start concerning safety-relevant information in front of the lead vehicle. Regarding the impact on road accident statistics, it must always be considered that it is impossible to anticipate all potential factors compromising road safety. Especially when introducing new technology, a close observation of a range of parameters is desirable. In the case of platooning, this is not only referring to the automated system but also to the temporarily inactive driver and the interaction with other road users. The impact on driver attention and effects of risk compensation or adverse behavioural changes of other road users in reaction to platoons, etc., cannot be ruled out yet. Either way, systematic collection of a wide range of safety-relevant indicators will be key for platooning and the safety of all road users involved.

References

1. Verordnung des Bundesministers für Verkehr. Innovation und Technologie über die Rahmenbedingungen für automatisiertes Fahren. Automatisiertes Fahren Verordnung, AutomatFahrV
2. Bundesgesetz vom 6. Juli 1960 (1960) mit den Vorschriften über die Straßenpolizei erlassen werden. Straßenverkehrsordnung 1960. StVO
3. Verordnung der Bundesministerin für Verkehr (2012) Innovation und Technologie über die Sicherung von Eisenbahnkreuzungen und das Verhalten bei der Annäherung an und beim Übersetzen von Eisenbahnkreuzungen. Eisenbahnkreuzungsverordnung 2012. EisbKrV
4. Energy Mobility Innovation Federal Ministry Climate Action (2020) Environment and technology. Testing of automated driving on public roads, code of practice
5. Jansen R, Lotan T, Winkelbauer M, Bärgman J, Kovaceva J, Donabauer M, Pommer A, Musicant O, Harel A, Wesseling S, Christoph M, van Nes N (2017) Time headway between cars and powered two-wheelers. Technical report
6. Statistik Austria (2020) Verkehrsunfallstatistik 2014–2018
7. Asare S, Chang J, Staples B (2010) Truck platooning early deployment assessment-independent evaluation performance measures for evaluating truck platooning field deployments. Technical report August. U.S. Department of Transportation, Washington, DC, 8 Aug 2020
8. Nadler B, Nadler F, Strnad B (2014) Road safety inspection (RSI) manual for conducting RSI (RSI) manual for conducting RSI. Technical report 038

Susanne Kaiser is a psychologist and researcher at the Austrian Road Safety Board (KFV) since 2013. Her work is focused on human aspects in vehicle automation, driver state assessment, aggression in traffic as well as traffic safety culture. Susanne sis a member of HUMANIST VCE and the German Society of Traffic Psychology while also teaching Human Factors in Mobility at the University of Applied Sciences FH Technikum Wien.

Martin Winkelbauer graduated in Mechanical Engineering at the Vienna Technical University and joined the Austrian Road Safety Board in 1993. After mainly working in the field of driver education and vehicle technology, he later expanded his expertise towards efficiency assessment of road safety measures. Today, he is a senior researcher managing and participating in research projects on national and international level. His focus is still on vehicle-related safety and driver education, but also cargo security, motorcycle safety, efficiency assessment and international cooperation.

Erwin Wannenmacher is a technical expert for traffic engineering at the Austrian Road Safety Board (KFV). His work is mainly focused on accident analysis, Road Safety Inspection, implementation of safety audits and safety in public spaces in national and international projects. He is a lecturer for the training of Road Safety Auditors and Inspectors as well as lighting technicians. Furthermore, he regularly participates in various working groups and committees to prepare technical regulations in the framework of the research community and road traffic (FSV).

Philipp Blass is a traffic safety researcher at the Austrian Road Safety Board (KFV) with a strong focus on automated driving and advanced driver assistance systems (ADAS).Within this field he contributes to several national and international research projects and aims to educate novice as well as experienced drivers in the topic of ADAS.

Hatun Atasayar is an urban planner and researcher at the Austrian Road Safety Board (KFV) with focus on traffic safety and transportation. She works in several national R&I projects on the aspects of the digitalisation of the transport system with a particular focus on C-ITS and automated mobility.

Chapter 14
Business Models, Economy and Innovation

Patrick Brandtner, Andrea Massimiani, Matthias Neubauer, Oliver Schauer, Wolfgang Schildorfer, and Gerold Wagner

Abstract Emerging technologies may trigger rethinking existing business models, clearly highlighting economic benefits and analysing effects on innovation systems. Truck platooning as one emerging technology in the area of road freight transport promises to improve efficiency and safety and requires different stakeholders (e.g. road operators, freight forwarders, truck manufacturers, etc.) to adapt their business models. In this chapter, key aspects when developing a truck platooning business model from a road operator's perspective will be summarised based on related work and interviews/workshops conducted in the Connecting Austria project. Furthermore, the relevance of ongoing trend monitoring to continuously adapting business models is discussed in this chapter and applied for logistics and automated driving within the Connection Austria consortium.

Keywords Trend monitoring · Truck platooning business model canvas · Truck platooning value proposition canvas

14.1 Key Aspects of a Truck Platooning Business Model from a Road Operator's Perspective

Truck platooning aims to increase efficiency in terms of traffic and fuel efficiency as well as traffic safety. As such, it represents a promising means for different stakeholders to improve future transport of goods. Related work targeted at business models for truck platooning investigates business models for different stakeholders, e.g. road operators, freight forwarders/logistics operators, truck manufacturers or platoon service provider. Freight forwarder will typically benefit from (cf. [9])

- Reduced fuel consumption and fuel costs.
- Reduced emissions and emission-related costs.

P. Brandtner · A. Massimiani · M. Neubauer · O. Schauer · W. Schildorfer (✉) · G. Wagner
Department of Logistics, University of Applied Sciences Upper Austria, Steyr, Austria
e-mail: wolfgang.schildorfer@fh-steyr.at

© The Author(s) 2022
A. Schirrer et al. (eds.), *Energy-Efficient and Semi-automated Truck Platooning*,
Lecture Notes in Intelligent Transportation and Infrastructure,
https://doi.org/10.1007/978-3-030-88682-0_14

- Reduced requirements for drivers when platooning might lead to adaptations of maximum driving times, thus affecting personnel costs and transport times.
- In higher automation levels, driver in following vehicles in a platoon may be allowed to rest; thus, shortening long-distance transport times may be supported.

However, gaining these benefits will depend on influence factors such as (cf. [1, 9, 11])

- Gain sharing model between trucks in a platoon.
- Pricing model of platoon service providers.
- Percentage of highway kilometres eligible for platooning and amount of viable platooning time.
- Penetration of trucks with platooning technology (SAE L1-L5).
- Ability to form multi-fleet, multi-brand platoons.
- National and European legal regulation for truck drivers' resting times or minimum distance between trucks in a platoon.
- Internalisation of external costs (e.g. air pollution) for logistics operators.

Platoon service providers (PSP) are considered to play a crucial role when it comes to the deployment of truck platooning. Central activities of PSPs are the platoon management and the financial clearing between platoon members. Examples for business model elements of truck platooning have already been published by [1, 5, 9, 11, 14]. In Connecting Austria, the support of truck platooning via C-ITS has been a vital element. Lu et al. (cf. [12]) consider C-ITS applications as important means to increase safety, efficiency and sustainability of road transport. They conducted a survey and revealed that respondents consider C-ITS services as important and relevant. However, the study also indicated a low willingness to pay, which demands for sustainable business models. Since C-ITS is an important issue for safe and efficient truck platooning and related work does not discuss business models in the C-ITS context, in addition to previous related work a business model canvas (cf. [15]) for truck platooning has been developed from the point of view of a road operator.

The result is depicted in Fig. 14.1. The results gained in Connecting Austria are based on related work and interviews/workshops, which are referenced using (CA) in Fig. 14.1 . In general, participants of empirical studies conducted in Connecting Austria provided an informed consent to participate and to publish the results. Furthermore, interview data was anonymised and securely stored internally in the University of Applied Sciences Upper Austria. In advance to the empirical studies, the authorisation of the involved parties related to the interviews was provided.

Based on the gathered data, a value proposition canvas (cf. [16]) has been developed for two customer segments. The value proposition canvas especially details the elements "value proposition" and "customer segment" of the business model canvas. The two customer segments investigated in detail are (1) the "government" as owner of public road operators (compare Fig. 14.2) and (2) the "platoon service providers". Since the information (e.g. dynamic information on platooning-eligible highway segments) provided by the road operator is highly relevant for platoon service providers when planning and managing platoons, a value proposition canvas for the customer segment "platoon service provider" has been developed (compare Fig. 14.3).

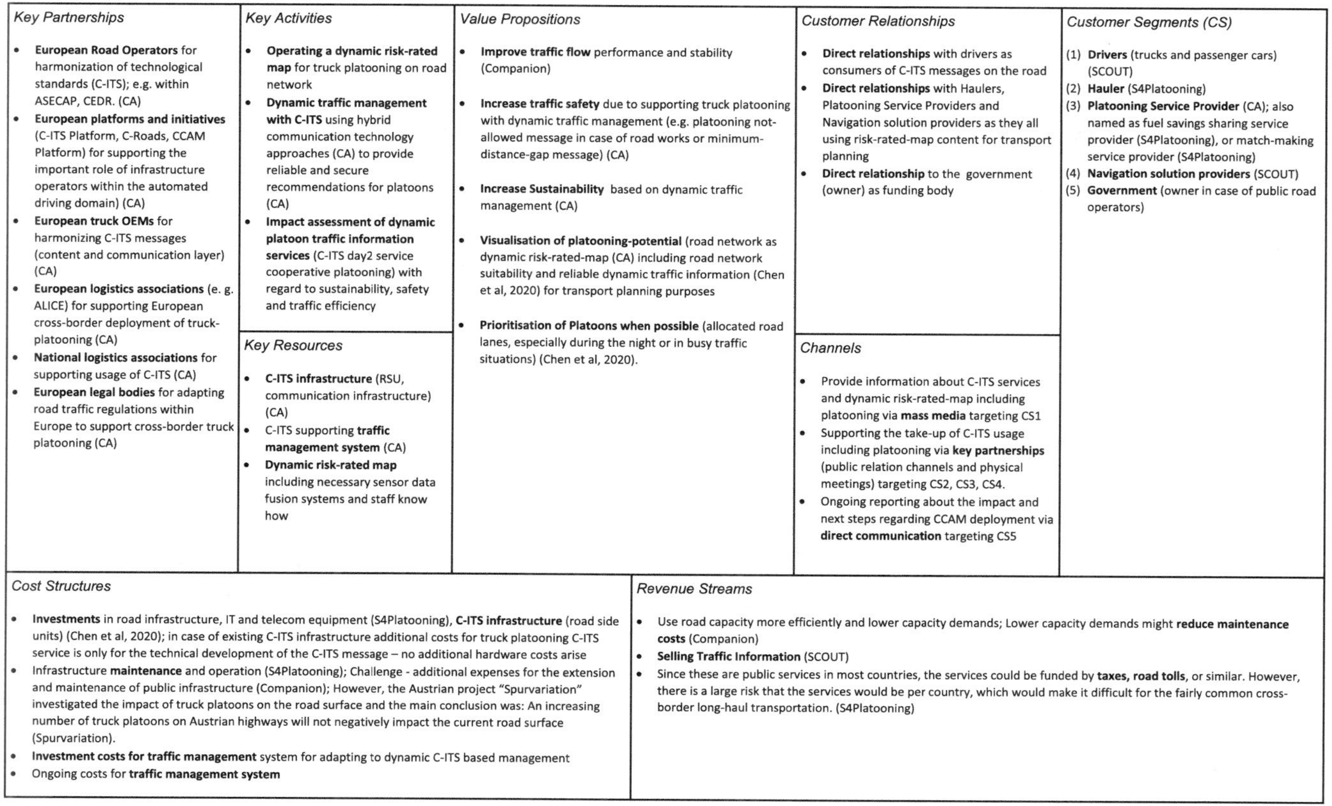

Key Partnerships
- **European Road Operators** for harmonization of technological standards (C-ITS); e.g. within ASECAP, CEDR. (CA)
- **European platforms and initiatives** (C-ITS Platform, C-Roads, CCAM Platform) for supporting the important role of infrastructure operators within the automated driving domain) (CA)
- **European truck OEMs** for harmonizing C-ITS messages (content and communication layer) (CA)
- **European logistics associations** (e.g. ALICE) for supporting European cross-border deployment of truck-platooning (CA)
- **National logistics associations** for supporting usage of C-ITS (CA)
- **European legal bodies** for adapting road traffic regulations within Europe to support cross-border truck platooning (CA)

Key Activities
- **Operating a dynamic risk-rated map** for truck platooning on road network
- **Dynamic traffic management with C-ITS** using hybrid communication technology approaches (CA) to provide reliable and secure recommendations for platoons (CA)
- **Impact assessment of dynamic platoon traffic information services** (C-ITS day2 service cooperative platooning) with regard to sustainability, safety and traffic efficiency

Key Resources
- **C-ITS infrastructure** (RSU, communication infrastructure) (CA)
- **C-ITS supporting traffic management system** (CA)
- **Dynamic risk-rated map** including necessary sensor data fusion systems and staff know how

Value Propositions
- **Improve traffic flow** performance and stability (Companion)
- **Increase traffic safety** due to supporting truck platooning with dynamic traffic management (e.g. platooning not-allowed message in case of road works or minimum-distance-gap message) (CA)
- **Increase Sustainability** based on dynamic traffic management (CA)
- **Visualisation of platooning-potential** (road network as dynamic risk-rated-map (CA) including road network suitability and reliable dynamic traffic information (Chen et al, 2020) for transport planning purposes
- **Prioritisation of Platoons when possible** (allocated road lanes, especially during the night or in busy traffic situations) (Chen et al, 2020).

Customer Relationships
- **Direct relationships** with drivers as consumers of C-ITS messages on the road
- **Direct relationships** with Haulers, Platooning Service Providers and Navigation solution providers as they all using risk-rated-map content for transport planning
- **Direct relationship** to the government (owner) as funding body

Channels
- Provide information about C-ITS services and dynamic risk-rated-map including platooning via **mass media** targeting CS1
- Supporting the take-up of C-ITS usage including platooning via **key partnerships** (public relation channels and physical meetings) targeting CS2, CS3, CS4.
- Ongoing reporting about the impact and next steps regarding CCAM deployment via **direct communication** targeting CS5

Customer Segments (CS)
(1) **Drivers** (trucks and passenger cars) (SCOUT)
(2) **Hauler** (S4Platooning)
(3) **Platooning Service Provider** (CA); also named as fuel savings sharing service provider (S4Platooning), or match-making service provider (S4Platooning)
(4) **Navigation solution providers** (SCOUT)
(5) **Government** (owner in case of public road operators)

Cost Structures
- **Investments** in road infrastructure, IT and telecom equipment (S4Platooning), **C-ITS infrastructure** (road side units) (Chen et al, 2020); in case of existing C-ITS infrastructure additional costs for truck platooning C-ITS service is only for the technical development of the C-ITS message – no additional hardware costs arise
- Infrastructure **maintenance** and operation (S4Platooning); Challenge – additional expenses for the extension and maintenance of public infrastructure (Companion); However, the Austrian project "Spurvariation" investigated the impact of truck platoons on the road surface and the main conclusion was: An increasing number of truck platoons on Austrian highways will not negatively impact the current road surface (Spurvariation).
- **Investment costs for traffic management** system for adapting to dynamic C-ITS based management
- Ongoing costs for **traffic management system**

Revenue Streams
- Use road capacity more efficiently and lower capacity demands; lower capacity demands might **reduce maintenance costs** (Companion)
- **Selling Traffic Information** (SCOUT)
- Since these are public services in most countries, the services could be funded by **taxes, road tolls**, or similar. However, there is a large risk that the services would be per country, which would make it difficult for the fairly common cross-border long-haul transportation. (S4platooning)

Fig. 14.1 Business model canvas for truck platooning from a road operator's point of view. www.strategyzer.com adapted from [15]. Courtesy of Strategyzer AG

Value Proposition Canvas for the Government

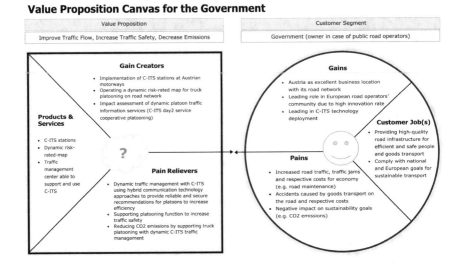

Fig. 14.2 Value proposition canvas for the customer segment government. www.strategyzer.com adapted from [16]. Courtesy of Strategyzer AG

Value Proposition Canvas for the Platoon Service Provider

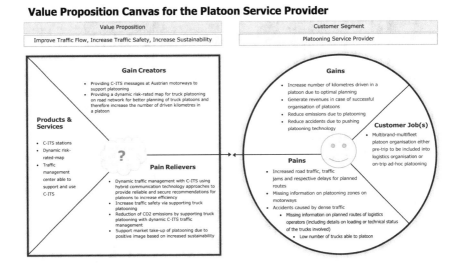

Fig. 14.3 Value proposition canvas for the customer segment platoon service provider. www.strategyzer.com adapted from [16]. Courtesy of Strategyzer AG

The results of analysing both value proposition canvas indicate:

- Commonalities within the customer segments.

  - With regard to **pains** while doing their job(s), both segments suffer from increased road traffic, traffic jams and the respective costs as well as accidents on the motorway and their negative implications.
  - **Gains** for both segments in doing their jobs are to reduce emissions and increase traffic safety due to supporting platooning technology.

- Differences within the customer segments.

  - **Customer Job(s)**: While the platooning service provider focuses on the multi-brand–multi-fleet platoon organisation either pretrip to be included into logistics organisation or on-trip ad hoc platooning, the government defines its jobs as providing high-quality road infrastructure for efficient and safe people and goods transport and complies with national and European goals for sustainable transport.
  - With regard to **pains** while doing their job(s), the differences are: for the platoon service provider, the missing information on planned routes of logistics operators (including details on loading or technical status of the trucks involved) and the low number of trucks able to platoon are the main pain points. For the government, the main pain point is the negative impact of increasing freight transport on sustainability goals (e.g. CO_2 emissions)
  - The **gains** in the customer segments differ as follows: the platooning service provider mainly focuses on the increasing number of kilometres driven in a platoon due to optimal planning and the generation of revenues in case of successful organisation of platoons. In comparison, the main gains for the government are to support Austria as excellent business location with its road network, foster the leading role in European road operators' community due to high innovation rate and enlarge the leading role in C-ITS technology deployment.

- Commonalities within the value proposition.

  - **Products and services** are the same for both customer segments.

    C-ITS stations.
    Dynamic risk-rated map.
    Traffic management centre able to support and use C-ITS.

  - With regard to **gain creators**, for both customer segments the first common issue is providing C-ITS messages at Austrian motorways to support platooning, while the second common gain creator is the provision of a dynamic risk-rated map.
  - The main commonalities with respect to **pain relievers** are

    Dynamic traffic management with C-ITS using hybrid communication technology approaches to provide reliable and secure recommendations for platoons to increase efficiency.
    Increase traffic safety via supporting truck platooning.

Reduction of CO_2 emissions by supporting truck platooning with dynamic C-ITS traffic management.

- Differences within the value proposition.

  - With regard to **gain creators**, there are two differences for the customer segments: only for the customer segment "government", the implementation of C-ITS stations as well as the C-ITS impact assessment is important.
  - One difference regarding **pain relievers** mentioned is the support of the market take-up of platooning due to positive image based on increased sustainability. Road operators proactive cooperation with truck manufacturers can support this issue and help the customer segment "platooning service provider".

So far, roadmaps for the deployment of truck platooning systems were presented by different institutions (cf. [5, 13]). Typically, the introduction of automated driving assistance function is structured along the automation levels (SAE L1-L4) and a time horizon. The deployment is considered to be evolutionary, i.e. incremental from one level to the next. To monitor and foster truck platooning deployment, not only vehicle innovations need to be investigated, but also innovations related to intelligent road infrastructures as well as the innovation readiness of logistics operators. For this reason, continuous trend monitoring is crucial to be prepared for future trends and to be able to adjust business models accordingly. Within the next section, trend monitoring as one main key feature for business model development and innovation is explained and applied in the context of logistics and automated driving.

14.2 Trend Monitoring as a Key Feature for Business Model Development and Innovation

14.2.1 Relevance of Trend Monitoring for Business Model Development

Markets, value networks and the organisational environment are changing faster than ever before. Once stable market mechanisms and predictable developments have become more and more instable and complex, developments and trends can hardly be predicted. With emerging topics and technologies such as artificial intelligence, cyber-physical systems, robotics or augmented and mixed reality, organisations have to deal with an increasing amount of strategic uncertainty. The ability to identify relevant trends and developments and to analyse their implications on current businesses becomes a crucial capability. Same applies to deriving opportunities for innovating existing and for generating new business models based on relevant trends (cf. [10]). Once reliable and proven business models may need to be innovated or completely rethought. Systematic and structured trend monitoring provides an approach of how to identify, analyse and continuously re-evaluate relevant trends and developments in the organisational environment. The outputs and findings of trend monitoring represent an important knowledge base for evaluating the applicability and future-fit of existing business models and for deriving potential business model innovation opportunities (cf. [6]). Existing literature shows that organisations with established trend monitoring initiatives can observe and track a larger amount of strategically relevant issues than those without such systems. Trend monitoring activities should not focus on external sources only, but also allow for integrating internal information, sources, networks and structures as well (cf. [2]). A comparison of practices and literature on trend monitoring shows that there are basically three steps involved in trend monitoring activities: (i) definition of search fields, (ii) identification and collection of inputs and trends and (iii) evaluation of trends and deduction of strategic implications (cf. [3]). In step 1, strategic innovation search fields provide the basis for all following activities and lay the foundation for trend identification and input collection (cf. [8]). Within defined search fields, input can systematically be collected and trends can be identified in step 2. Finally, the collected and clearly described trends and developments can be used as the basis for trend evaluation in the third step (Brandtner, 2018). Typical evaluation criteria may include the maturity or the probability of a certain trend and the estimated impact of the trend on the organisation's businesses (cf. [17]).

14.2.2   Applying Trend Monitoring in the Context of Logistics and Automated Driving

Based on the three-step logic of trend monitoring discussed in the previous section, a model for applying trend monitoring in the context of logistics and automated driving was developed. Fig. 14.4 depicts the developed approach:

In step one, a group of researchers and professors in the area of transport logistics, automated driving and strategic foresight respectively trend monitoring defined search terms. This approach ensured the consideration of a complete and domain-based set of keywords and search fields for the subsequent stages of the procedure. Step 2—identification and collection of inputs and trends—was headed by a research associate with expertise in trend monitoring and applying the software TRENDONE. In the third step—evaluation of trends and deduction of strategic implications—the prepared trend profiles were sent out for evaluation to domain experts. In total, 12 experts from the sectors of smart mobility, logistics and transport, research and academia, technology development, public domain, software development and R&D evaluated the trend profiles. All the experts provided an informed consent to participate in the study and to publish the results. Based on the expert evaluations, an aggregated view of the trend profiles was generated and visualised in the form of a trend radar (compare Fig. 14.5).

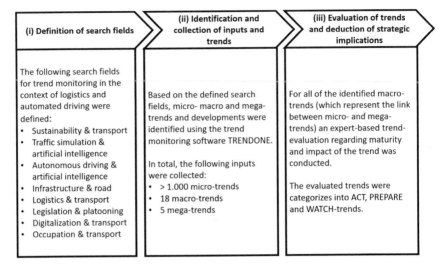

(i) Definition of search fields	(ii) Identification and collection of inputs and trends	(iii) Evaluation of trends and deduction of strategic implications
The following search fields for trend monitoring in the context of logistics and automated driving were defined: • Sustainability & transport • Traffic simulation & artificial intelligence • Autonomous driving & artificial intelligence • Infrastructure & road • Logistics & transport • Legislation & platooning • Digitalization & transport • Occupation & transport	Based on the defined search fields, micro- macro and mega-trends and developments were identified using the trend monitoring software TRENDONE. In total, the following inputs were collected: • > 1.000 micro-trends • 18 macro-trends • 5 mega-trends	For all of the identified macro-trends (which represent the link between micro- and mega-trends) an expert-based trend-evaluation regarding maturity and impact of the trend was conducted. The evaluated trends were categorizes into ACT, PREPARE and WATCH-trends.

Fig. 14.4 Developed trend monitoring approach for logistics and automated driving

14.2.3 Implications for Business Model Development Related to Logistics and Automated Driving

The results of the trend evaluation and the depiction of these trends on the developed trend radar showed that:

- 5 macro-trends were categorised as ACT trends.
- 11 macro-trends fell into the PREPARE category.
- 2 of the selected macro-trends were WATCH trends.
- None of the selected macro-trends were evaluated as OUT OF SCOPE trend, indicating that the search fields were defined in a focused and matching way.

The final trend radar (compare Fig. 14.5) looks as follows and represents the main outcome of the conducted trend monitoring approach. ACT trends have a high potential of influencing logistics and automated driving already now or at least in the very near future. They show a high degree of maturity and are tangible and applicable in the given context. In the defined field of logistics and automated driving, the participants assigned five trends to the ACT category (cf. Table 14.1). For these ACT trends, the recommendation for action is to integrate them into the current innovation portfolio and start defining concrete actions in these areas. PREPARE trends have a high potential of influencing logistics and automated driving. However, their maturity is not as far progressed as it is the case for ACT trends. Hence, organisations may still have some time to deal with these trends and prepare a strategy on how they get further information. In the given context, 11 trends were identified as PREPARE trends (cf. Table 14.1). Finally, WATCH trends are those trends that currently show little potential of influencing automated driving within the logistics domain but have

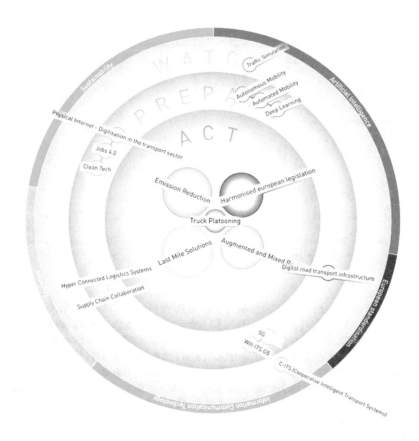

Fig. 14.5 Trend radar for logistics and automated driving

a high degree of maturity. These trends should be watched; i.e. organisation should continuously monitor the developments of these trends also in other industries in order not to miss relevant changes. In the given context, 2 trends were evaluated as WATCH trends (cf. Table 14.1).

14.3   Discussion and Conclusion

A main barrier for innovation and the related investment may represent a technological lock-in effect (cf. [7]). High investment cost in C-ITS infrastructure for road operators has been such a barrier the last years. On a European level, the C-ITS Deployment Platform was set up in 2014 to support a harmonised deployment across European countries (cf. [12]). With ASFINAG's roll-out of C-ITS stations at 2000

Table 14.1 Evaluation results of trend monitoring for automated driving within the logistics domain

ACT	Prepare	Watch
Emission reduction	Autonomous driving	Traffic simulation
Last mile solutions	Automated mobility	C-ITS
Truck platooning	Deep learning	
Harmonised European legislation	5G	
Augmented and mixed reality	P-WiFi (ITS-G5)	
	Digital road infrastructure	
	Clean tech	
	Jobs 4.0	
	Supply chain collaboration	
	Hyperconnected logistics	
	Physical Internet	

highway kilometres in November 2020, the C-ITS deployment started in Austria and the foundation for Day-1 and Day-2 C-ITS services like cooperative platooning was established. Even if the initial investment in Austria took already place, viable business models for the operation, maintenance and the enhancement of C-ITS services are still an open issue. Examples in existing publications propose as revenue streams taxes/tolls, reduced road maintenance costs or earnings from data provision services. However, when developing/deploying new C-ITS services viable business models should be investigated to support road operators and technology providers in taking informed decisions and sustainably deploying C-ITS services.

In addition to the C-ITS business model investigation, the relevance of ongoing trend monitoring to continuously adapting business models was discussed in this contribution. Doing so, a trend monitor related to logistics and automated driving was set up and rated by members of the Connecting Austria consortium. The members identified the following 5 ACT trends—emission reduction, last mile solutions, truck platooning, harmonisation of legislation and augmented and mixed reality. Emission reduction gained momentum in the last decades (cf. European Green Deal) and forces car manufacturers to rethink traditional propulsion systems, logistics operators to optimise transport processes and evaluate solutions for sustainable logistics operations, and road operators to increase traffic efficiency and enforce access regulations (e.g. Urban Access Regulations for Low Emissions Zones). Emission reduction is also crucial in urban areas and demands for innovative and sustainable last mile solutions (delivery robots, pickup stations, etc.). However, also long-distance transport between urban areas needs to be optimised. Truck platooning as such represents a means to increase efficiency and safety. A major prerequisite for truck platooning and automated driving is a harmonised legislation. Especially, long-distance transport across borders with truck platoons requires a harmonisation of legislation in order to support efficient operation and trigger platooning technology providers to offer products. As fifth trend, augmented and mixed reality has been identified as ACT trend. Augmented reality may support truck drivers via displaying relevant driving information (e.g. navigation information, display of C-ITS information). Overall, 11 PREPARE trends have been identified. These trends may be allocated to the following three clusters—automated driving technologies, clean tech and logistics trends. Automated vehicles (SAE L3-L5) are currently under development. Artificial intelligence algorithms (deep learning) and communication technologies such as 5G and ITS-G5 fuel automated driving functions. Furthermore, information provided from the road infrastructure related to the available infrastructure support for automated driving functions (cf. ISAD levels [4]) will be relevant to implement highly automated driving in a safe manner. Logistics trends will adopt new technologies and aim to apply them in order to meet sustainability goals. Furthermore, logistics operators will need to rethink current job profiles especially when it comes to highly automated driving. Clean tech, as third cluster, comprises renewable energy technologies, technologies to reduce emissions, safe energy and resource exploitation.

Finally, two trends—traffic simulation and cooperative intelligent transport systems (C-ITS)—were assigned to the WATCH category. This category typically comprises trends that have a high maturity. However, the trend potential is currently considered to be low or not yet adopted within a given domain. Regarding the judgements of the trend monitor participants, the potential of C-ITS and traffic simulation for automated driving within the logistics domain needs to be monitored, assessed and adopted in future.

References

1. Axelsson J, Bergh T, Mårdberg B, Johansson A, Svenson P, Åkesson V (2020) Truck platooning business case analysis. In: RISE Rapport. Available from: http://urn.kb.se/resolve?urn=urn:nbn:se:ri:diva-44282
2. Brandtner P (2017) Design and evaluation of a process model for the early stages of product innovation. In: New waves in innovation management research - ispim insights: series in innovation studies, pp 149–162
3. Brandtner P, Gaubinger K, Auinger A, Markus Helfert, Rabl M (2014) Dealing with uncertainty in innovation management-an empirical analysis of activities and method use in innovative organizations. In: ISPIM Conference Proceedings, pp 1. The International Society for Professional Innovation Management (ISPIM)
4. Carreras A, Daura X, Erhart J, Ruehrup S (2018) Road infrastructure support levels for automated driving. In: Proceedings of the 25th ITS World Congress, Copenhagen, Denmark, pp 17–21
5. Dubbert J, Wilsch B, Zachäus C, Meyer G (2018) Roadmap for accelerated innovation in level 4/5 connected and automated driving. In: International forum on advanced microsystems for automotive applications. Springer, Berlin, pp 183–194
6. Fergnani A, Hines A, Lanteri A, Esposito M (2020) Corporate foresight in an ever-turbulent era. In: European business review, pp 6–33
7. Foxon TJ (2014) Technological lock-in and the role of innovation. In: Handbook of sustainable development. Edward Elgar Publishing
8. Groher W, Rademacher F-W, Csillaghy A (2019) Leveraging AI-based decision support for opportunity analysis. Technol Innov Manag Rev 9(12)

9. Hanelt A, Leonhardt D, Hildebrandt B, Piccinini E, Kolbe L (2019) Pushing and pulling – digital business model innovation and dynamic capabilities. In: Journal of competences, strategy & management, vol. 10, pp 55–78
10. Højland J, Rohrbeck R (2018) The role of corporate foresight in exploring new markets-evidence from 3 case studies in the bop markets. Technol Anal Strateg Manag 30(6):734–746
11. Janssen R, Zwijnenberg H, Blankers I, de Kruijff J (2015) Truck platooning–driving the future of transportation. In: TNO: The Hague. Available from: https://repository.tno.nl/islandora/object/uuid%3A778397eb-59d3-4d23-9185-511385b91509
12. Lu M, Türetken O, Adali OE, Castells J, Blokpoel R, Grefen P.W.P.J (2018) C-its (cooperative intelligent transport systems) deployment in Europe: challenges and key findings. In: 25th ITS world congress, pp EU–TP1076
13. Meyer G (2019) European roadmaps, programs, and projects for innovation in connected and automated road transport. In: Road vehicle automation, vol 5. Springer, Berlin, pp 27–39
14. Muratori M, Holden J, Lammert M, Duran A, Young S, Gonder J (2017) Potentials for platooning in us highway freight transport. Technical report, National Renewable Energy Lab.(NREL), Golden, CO (United States)
15. Osterwalder A, Pigneur Y (2010) Business model generation: a handbook for visionaries, game changers, and challengers. Wiley
16. Osterwalder A, Pigneur Y, Bernarda G, Smith A (2014) Value proposition design: how to create products and services customers want. Wiley
17. Pierenkemper C, Gausemeier J (2020) Developing strategies for digital transformation in smes with maturity models. In: ISPIM conference proceedings. The International Society for Professional Innovation Management (ISPIM), pp 1–20

Patrick Brandtner is a professor of supply chain management at the University of Applied Sciences Upper Austria. His research interests cover strategic foresight, data analytics and innovation management in the context of supply chain management. Patrick has been involved in several national and international research projects and teaches in the master's programs supply chain management as well as digital transport and logistics management classes such as innovation methods in logistics and transport, data analytics or design thinking.

Andrea Massimiani born 1975, studied, after her active career as a competitive athlete, economics at the Johannes Kepler University of Linz and gained extensive experience in banking and industry. She is a research associate at the University of Applied Sciences Upper Austria, focusing on trend monitoring, corporate foresight and retail logistics.

Matthias Neubauer is a professor at the University of Applied Sciences Upper Austria for logistics information systems. His research interests cover human-computer interaction, intelligent transportation systems as well as cooperative, connected and automated mobility. Matthias is involved in international and national research projects and teaches in the master's program digital transport and logistics management classes such as process management, distributed logistics systems or geoinformation systems.

Oliver Schauer works as a professor in transport logistics and mobility at the University of Applied Sciences Upper Austria, with emphasis on the fields of sustainable logistics, Physical Internet and digital transformation. He is also head of studies of the master's program "Digital Transport- and Logistics-Management". Prior to his actual profession Oliver gained strong experience in the executive management of leading Austrian logistics providers.

Wolfgang Schildorfer is a person who likes the road he still walks to find new chances. Since October 2018, he has been Professor for Transport Logistics and Mobility at the University of Applied Sciences Upper Austria. His research focus is on innovation, business models and evaluation in transport logistics, smart hyperconnected logistics systems, (urban) mobility, sustainable transport systems and new technology markets (C-ITS, CCAM, automated driving, truck platooning).

Gerold Wagner is a professor for business informatics at the University of Applied Sciences Upper Austria. His research interests cover business information systems, new business models and E-Learning. Gerold is coordinator of the master's program supply chain management and is involved in national and international research projects.

Chapter 15
Advanced Powertrain Systems for Platooning-Capable Trucks

Michael Nöst, Christian Doppler, and Alexander Mladek

Abstract This chapter deals with the interaction of platooning-capable trucks and their powertrain systems. In a first step, prospective propulsion systems and their characteristics for platooning are discussed. Therefore, different topologies are analysed, also in terms of the intended use cases. These considerations are made mainly according to CO_2-limitation efforts in the background. Secondly, thermal management regarding platooning is in the focus. Investigations on the influence of the air mass flow for an internal combustion engine (ICE) operated truck are presented. Further, thermal management challenges in combination with a fuel cell operated truck are discussed. For this purpose, dedicated solutions and methods in development are presented. Finally, essential future research fields are outlined.

15.1 Introduction

Possible effects of truck platooning, especially on propulsion systems, are being discussed in this chapter. Improving vehicle performance is one of the key issues for reducing the environmental footprint of long-distance freight transport. Measures oriented to reduce emissions of CO_2 and pollutants can be classified in two blocks, namely *propulsion system* and *vehicle design*.

Propulsion System: Beside the optimisation and adaptation of ICE-driven powertrains for renewable fuels, another trend is the increasing rate of electrified vehicles like battery electric vehicles (BEVs), hybridised electric vehicles (HEVs) or plug-in hybridised electric vehicles (PHEVs), including the electric energy storage not only using batteries, but also from fuel cell or from high-efficient ICE energy converters running on hydrogen, as well as electric road systems.

Vehicle Design: The main parameters are the aerodynamics and the rolling resistance. The optimisation of the aerodynamics is today restricted by regulatory con-

M. Nöst (✉)
IESTA—Institute for Advanced Energy Systems and Transport Applications, Graz, Austria
e-mail: michael.noest@iesta.at

C. Doppler · A. Mladek
Virtual Vehicle Research GmbH, Graz, Austria

A. Schirrer et al. (eds.), *Energy-Efficient and Semi-automated Truck Platooning*,
Lecture Notes in Intelligent Transportation and Infrastructure,
https://doi.org/10.1007/978-3-030-88682-0_15

straints, which are likely to be relaxed soon (e.g. platooning). The topic aerodynamic advantages due to platooning is discussed in detail in Chaps. 5 and 6. Combining automated driving and electrified propulsion systems can substantially improve the road freight transport efficiency, although high costs of the battery electric heavy-duty vehicles (HDVs) are hindering their adoption by transportation companies.

15.2 CO$_2$ Emission Reduction by Different Application Domains

[1]Compared to 1990, domestic greenhouse gas emissions in the EU are to be reduced by at least 40% until 2030. All sectors have to make their contribution. The road transport sector is particularly essential for reducing greenhouse gas emissions and decarbonising the economy in the EU [1]. Road freight transport is essential for the development of trade on the European continent. Trucks carry around 70% of freight transported over land. Emissions from HDVs (trucks, buses and coaches) account for about 6% of total EU CO$_2$ emissions and 25% of road CO$_2$ emissions in the EU. Without any further action, CO$_2$ emissions caused by HDVs are projected to grow by 9% over the period 2010 to 2030 due to the increasing transportation activities. Since 2018, there has been a legislative proposal for CO$_2$ emission standards for HDVs in the EU. In order to reach the proposed objectives for the new vehicles coming into the market, truck OEMs should most likely manage a fuel consumption decrease by more than 3% per year until 2030.

According to EUROSTAT 2018, about 80% of all freight transport (in terms of tonne kilometre) is realised on long haul (over a distance of 150 km or more).

The needs are obviously different by vehicle usage and application domain. High technologies will gradually be extended to different use cases, from simple to increasingly complex environments: the European Road Transport Research Advisory Council (ERTRAC) has defined various areas of application for trucks, classified from simple environment to very complex (from the challenge for an automated operation point of view) as follows [2]:

- *"Confined area:"* Private area, terminals, ports ⇒ simple environment.
- *"Hub-To-Hub:"* Partly public road, from companies to ports or terminals ⇒ relatively simple environment.
- *"Highways:"* Public roads and highways ⇒ complex environment.
- *"Urban environment:"* Cities and public roads ⇒ very complex environment.

As also stated in [2] the *highway* application is mainly responsible for the major part of the CO$_2$ emissions. The *urban environment* or *confined areas* have less significant impact on the share of CO$_2$ emissions; therefore, we focus the following discussion on the area of *highways*.

---

[1] Some text in this section is adapted from [6] © 2014 ERTRAC. Reused with the permission of ERTRAC.

15.3 Ultra-low Emissions on Highways and Zero Emissions in Cities

[2]ERTRAC stated in [2] that potential and applicability of alternative propulsion technologies and fuels varies depending on different factors and the characteristic driving profile of a HDV. By 2030 and beyond, BEVs, requiring local charging stations, will be more suitable for short and medium distances. For long distances, hybridised powertrains with ICEs running on sustainable low-emission liquid or gaseous fuels, or alternatively electrified fuel cells, will be more suitable.

In February 2020, IVECO and FPT Industrial announced the plan to form a joint venture to equip Class 8 heavy-duty trucks through fuel cell technology.

The electricity sector is not yet fully decarbonised, though this is of course the goal in a far-future scenario. Energy storage, sourcing of battery materials and grid balancing might be the biggest practical issues at the time of increasing demand for electricity. Hydrogen is an energy carrier. Despite hydrogen being largely present in nature, it is not available as a pure element, so it must be produced using other sources of energy. The life cycle GHG emissions of the whole value chain (feedstock and energy) has to be considered.

15.4 Get the Right Infrastructure for Vehicle Energy Supply

A key ingredient of decarbonisation of future transport systems will be the availability of electrified vehicles including hybridised powertrains with ultra-low emission ICEs powered with low-carbon fuels. The development of both depends upon the rapid growth of an energy distribution network. This is relatively easy for liquid biofuels and liquid e-fuels, as it requires minor adaptation of the existing one. It is more challenging for gaseous low-carbon fuels (biogas, H_2), due to the cost of the refill stations. And, it is also challenging for electricity, due to its strong impact on the electricity grid and power stations that need to be implemented in size (with appropriate design guidelines and policy for making charging stations accessible for trucks) and electric power, to cover fast-charging energy needs.

---

[2] Some text in this section is adapted from [6] © 2014 ERTRAC. Reused with the permission of ERTRAC.

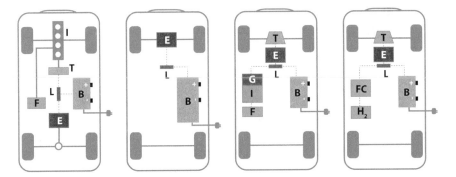

Fig. 15.1 Truck propulsion topologies, from left to right: **a** PHEV, **b** BEV, **c** EV + REX, **d** FCEV; B...battery, E...E.-motor, F...fuel tank, H_2...hydrogen tank, G...generator, I...internal combustion engine, L...power-split, T...Transmission, FC...Fuel cell

15.5 Different Topologies for Truck Drives

The transition to a more sustainable freight transportation sector requires the widespread adoption of electric vehicles powered by batteries (BEVs) or fuel cells (FCEVs) beside the possibility to hybridise powertrains with plug-in functionality (PHEVs).

Figure 15.1 shows schematically different propulsion topologies—relevant for platoon able trucks—explaining how different kind of power sources are linked to the drive train. (a) shows a typically PHEV in a parallel application with plug-in functionality, (b) shows a pure BEV (battery-driven electric vehicle) (c) a serial hybrid with ICE as range extender and (d) a fuel cell EV.

In (a), the engine as well as the electric machine is each connected to the front and rear axle. The parallel configuration shown in (a) is characterized by a higher transmission efficiency due to the mechanical link between ICE and the electric motor.

An advantage of a series hybrid—shown in (c)—is that the engine operates at its maximum efficiency point(s) thanks to the buffering of excess power; however, one disadvantage is the relatively low transmission efficiency at relatively high vehicle loads compared to other hybrid configurations. If the vehicle does not mainly drive in urban traffic outside the urban area, relatively large electric machines (kW) are needed to provide power for high vehicle speeds. In (d), the ICE is replaced through a fuel cell, used as a primary power source. Topologies (c) and (d) require less battery capacity than a pure BEV, visualised in topology (b).

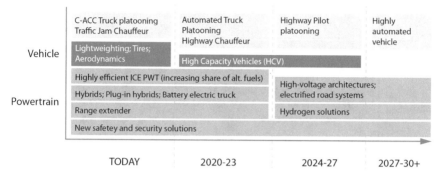

Fig. 15.2 Roadmap of possible technologies for highway domain

15.5.1   Truck Propulsion Systems for Highway Domain

Internal combustion engine propulsion systems are expected to remain the short- and mid-term technology of choice for long-distance intercity freight transportation. This will require continued focus on low emissions and very high energy conversion efficiency. Possible measures as renewable fuels, waste heat recovery and hybridisation should be seen in a short/mid-term time range, electric battery/fuel cells systems are the key technologies for the long-term range. Regarding vehicle energy density, for ICE-electric or fuel cell electric solutions, hydrogen offers a better alternative than electric batteries. But production efficiency of hydrogen and re-fuelling/charging infrastructure development need is also to be addressed, see Fig. 15.2.

15.5.2   Truck Propulsion Systems for Urban Domain

In future, short and medium distances may be more fitted for electric batteries or alternative powertrains to reduce NO_x, particle and noise emissions. Fully electric vehicles, improvements in battery technology, reduction of cost, mass and improvement life cycle impact are essential for the market up-take of electrified HDV for urban use.

For the different application domains, possible propulsion systems are discussed as an example for highway and urban domain, see Fig. 15.3.

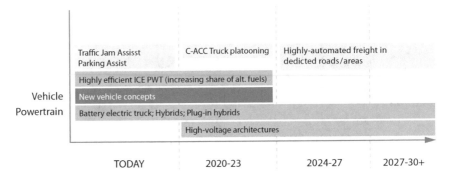

Fig. 15.3 Roadmap of possible technologies for urban domain

15.6 Importance of Thermal Management Concepts for Truck Drives

15.6.1 Motivation

HDV manufacturers constantly work on improving the fuel efficiency of the vehicle. This may be achieved on the engine side by a more efficient combustion process or at the chassis by increasing the aerodynamic parameters. Significant gains in the order of 5–10% can also be achieved by platooning, thus when one HDV travels in the slipstream of another HDV. However, the engine of an HDV needs to be cooled by the airflow through the engine compartment. This air mass flow through the radiator is significantly reduced while driving in a platoon for the following trucks. This may cause an insufficient cooling of the engine and may result in an engine failure [3]. To investigate the thermal behaviour of a platoon's vehicle, Connecting Austria project partners made tests on a closed testing area. The second HDV of a platoon consisting of three vehicles was equipped with sensors to measure coolant inlet and outlet temperatures, heat exchange and pressure loss of the cooling circuit.

15.6.2 Materials and Methods

The automotive proving ground in Zalazone, Hungary, [4], provides ample possibilities for dynamic on road tests of vehicles. It consists of numerous test tracks designed for a wide variety of purposes. Similar to the first measurement campaign, the second tests were performed at ZalaZone's handling course as well. This campaign incorporates additional tests related to the thermal behaviour of the cooling system of a truck in a platoon formation. The truck configuration consisted of two HDV (Volvo FH 540). An engine load trailer was additionally attached to the second HDV to

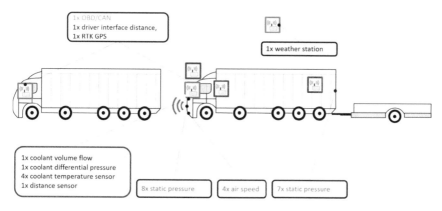

Fig. 15.4 Platoon sensors and load trailer for thermodynamic tests conducted at ZalaZone proving grounds

increase/vary the engine load depending on the test. The platoon was equipped with different sensors which are depicted in Fig. 15.4. For more details, see Sect. 6.2.3.

As described in Sect. 6.2.3, the monitoring of the cooling system of the truck was realised by a coolant volume flow sensor, a coolant differential pressure sensor as well as coolant temperature sensors at the inlet and outlet of the radiator. Also, the speed of the coolant air was measured at four different positions. In addition, the fan speed was recorded.

A laser distance sensor was installed at the front of the second HDV. Wind speed and wind direction, ambient air temperature and barometric pressure were measured at the weather station located some hundreds metres nearby the test section. Data was collected by highly synchronised data acquisition systems and transmitted via WLAN to a master data acquisition system. To increase the engine load, a trailer dynamometer (manufacturer: Unsinn) was used. The braking power was set' to 350 kW.

The distance between the two HDV was set to 7, 11 and 22 m. Each measurement consisted of monitoring the thermal parameters for several laps of the test circuit. The vehicle speed was set to 80 km/h.

15.6.3   Results

The coolant inlet temperature is the most important parameter for the assessment of the cooling circuit. The mean inlet temperature for test scenarios of a vehicle distance of 7 m (inset a.), 11 m (inset b.) and 22 m (inset c.) is shown in Fig. 15.5. For a vehicle distance of 11 m and 22 m, the inlet temperature was around 50°C at the start, increased to 92 °C at the second lap and remained stable at this temperature level for the remaining laps.

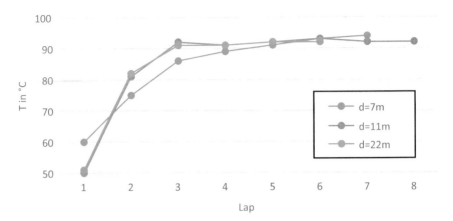

Fig. 15.5 Inlet temperature of the coolant for different vehicle distances

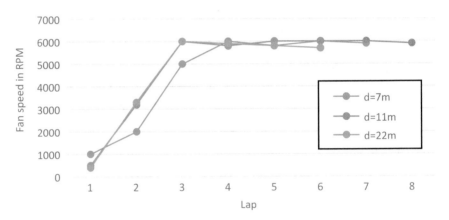

Fig. 15.6 Fan speed for different vehicle distances

However, the coolant inlet temperature at a vehicle distance of 7 m showed a different behaviour. The starting temperature was 60 °C, increased to 91 °C at the fourth lap but never converged to a stable temperature. There is still an increase of the temperature from the fifth to the sixth lap. Thus, there is an indication that the cooling system fails at very close platooning under a heavy load. In case of an insufficient heat exchange, the radiator fan has to increase the air mass flow. The results are shown in Fig. 15.6. The rotational speed of the fan increased to the maximum of 6000 rpm at the second lap for vehicle distances of 11 and 22 m and remains at this level for the remaining laps.

15.6.4   Discussion

Platooning leads to a reduced fuel consumption as the drag coefficient is significantly reduced in the slipstream. However, this also leads to a reduced radiator air mass flow, a reduced heat exchange and may result in a malfunctioning of the engine's cooling circuit. Measurement at the proving ground in ZalaZone indicated that for a vehicle distance of 7 m under a friction load of 350 kW the radiator inlet temperature does not converge. For larger vehicle distances a stable coolant radiator inlet temperature was reached. For all three distances, the radiator fan was operated at maximum speed. The operation of the fan may consume a part of the fuel consumption gain. Additionally, it has to be investigated if the radiator fan is designed for long operation under maximum speed.

15.7   Cooling Concepts on the Example of H_2 Driven Trucks

Due to the lack of air flow at the front heat exchanger with platooning, higher effort must be put in cooling the propulsion system. Especially when looking towards combustion engine and fuel cell engine, improved cooling performance needs to be ensured. In case of a pure electric propulsion system, heat fluxes for dissipation to the surrounding are much lower and less challenging. As nowadays a pure electric-operated HDV truck is not economic due to high battery weight, a serial hybrid concept is most promising. This comprises of an electric motor mechanically connected to the wheel's axes. The battery can be laid out, for example 100km pure electric driving (inner city). The major energy is provided by the range extender, either an ICE, operated in its sweet spot or by a fuel cell. With respect to emissions, the fuel cell is seen as the favourable solution for the near future.

   In Fig. 15.7, the Sankey diagram for heat flow of a fuel cell is shown. Approximately, 33% of the H_2 energy are dissipated to the ambient, while due to low temperature difference, a high constant mass flow must be ensured. Therefore, a bigger heat exchanger area and additional fan power is required. As platooning is contrary to this requirements, additional effort must be made.

   The concept in Fig. 15.8 is to be seen as the most common approach. This system is also called a two-circuit system as there is one low-temperature (LT) cycle for the e-motor and the power electronics, whereat especially the latter is in need of lower coolant temperatures which does not allow any satisfying use of one overall circuit architecture. The second circuit serves for cooling the stack and is called high-temperature (HT) circuit. The system is driven by an electrically controlled pump, and a particle filter can be used to avoid fouling of the PEMFC and the HX. Then, the temperature control of the humidifiers is common with the release of heat prior absorbed form the FC. In a next step, the compressor, respectively, the intercooler (if existing), for the reactant air is conditioned. Via convection at the radiator, the heat is released to the ambient air. Subsequently, an ion exchanger is used in case of no use

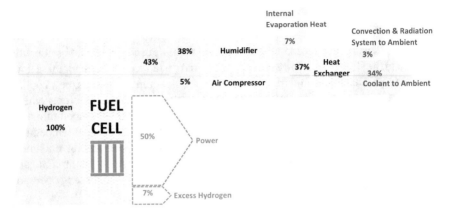

Fig. 15.7 Sankey diagram for heat flow of a fuel cell

of glycol within the cooling circuit. This assembly ensures to reduce, i.e. soda-ions, and restock the water with potassium ions. For this reason, the water loses most of its electrical conductibility.

The concept in Fig. 15.9 shows an approach for increasing the cooling efficiency of the HT-cooling circuit of the FC. By the use of an additional heat exchanger between HT- and LT circuit, respectively, HT- and A/C-circuit extra cooling performance can be gained. On the one hand, this goes along with higher (and for FC very important) cooling performances, and on the other hand, this leads to an increase of the total system efficiency because of better utilisation of the proper cooling capacities. It is required to execute vast prior investigations towards control and operation strategies in order to have a resilient validation for the temperatures in the cooling/conditioning cycle. The heat exchanger should be mounted in an exposed location (in front of the wheels), because of higher air flows at this position.

15.8 Outlook

To ensure sufficient cooling performance, *active cooling with a refrigerant system* as discussed above should be used. Therewith, the thermal drawback of platooning can be fully eliminated. As shown above, this approach requires a complex and big refrigerant system. In terms of future refrigerants with a low global warming potential, also propane can be investigated. Due to a current safety limit of 150g propane for refrigerant circuits, a compact refrigerant unit (CRU) should be considered. Such a system can significantly help to improve thermal efficiency of the powertrain without a big change of the vehicle architecture. Such a prototype was developed by virtual vehicle in the EU-project *Optemus* and is currently in research and development for series production, also for heavy-duty applications [5].

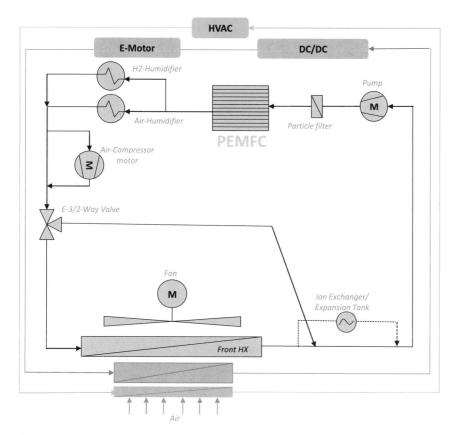

Fig. 15.8 Two-circuit cooling system including low-temperature (LT) and high-temperature (HT) circuits

Another methodology is to implement a *predictive thermal control*. With the help of the predicted road profile in a virtual model, the future heat fluxes are estimated and the thermal system behaviour for the next horizon can be estimated. Therefore, countermeasures to reduce the temperature can be already started at a very early stage, and high temperature peaks are avoided. In general, also lower auxiliary's energy consumption can be achieved with this approach due to operating at higher coolant temperatures, respectively, at better aggregates' performance points, see [6]. With prediction, the optimal time for driving in the platoon is estimated and the platoon can be selectively dissolved if too high thermal loads occur. In the case of upcoming hill climbs or city parts, for example, a dissolution much in advance to reaching temperature thresholds is started. In turn, at "normal" highway drives, platooning can be kept up long and potentially even at higher threshold temperatures as safety margins can be reduced.

Beside the described predictive thermal management, a predictive energy management (with focus on the operation of powertrain components) needs to be pursued.

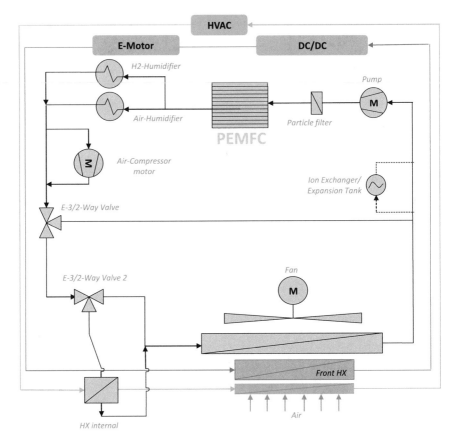

Fig. 15.9 Two-circuit cooling system with an additional heat exchanger connecting the HT and LT circuit, increasing the cooling efficiency of the HT-cooling circuit for the FC

It allows HDV applications in platooning to increase the efficiency of the power source system (fuel cell and battery). Further, this serves to mitigate the degradation of the components and to increase its lifetime. Also, when looking towards the state of charge of the battery, predictive measures will ensure to sustain battery charge as far as possible during platooning. With the focus on fuel cell electric vehicles (FCEV), future research focus for cooling is needed, especially due to the following facts [6, 7]:

- Reduction of large radiators cooling surfaces.
- Simplification of thermal management and reduction of parallel circuits with different temperature levels.
- Focus for predictive strategies towards uphill and platoon sections that have significant heat losses.

- The degradation of fuel cell modules is increased by frequent starts and shutdowns. Therefore, predictive strategies for activation and shutdown of fuel cell modules are required.

Due to the growing demand of transportation of goods, implementing only one of these measures discussed will not be sufficient enough to significantly increase fuel consumption efficiency and therefore reducing the environmental impact [8]. Therefore, the combination of automated driving and electrified propulsion systems will substantially improve the environmental impact and road freight transport efficiency. More details can be found in [9] in section *"Advanced Vehicle Concepts 2020+"*, subsection *"Research Requirements for Digitalization and Automatization of Vehicles and Infrastructure"*.

References

1. Eurostat (2019) Energy, transport and environment statistics. Technical report
2. ERTRAC (2019) Long distance freight transport a roadmap for system integration of road transport, pp 1–44
3. Block B, Huynh B, Boyle S, Stockar S, Geyer S, Li J, Huber J (2019) Analysis of the effect of vehicle platooning on the optimal control of a heavy duty engine thermal system, vol 4
4. Zalazone—automated proving ground. https://zalazone.hu/en/. 10 Jan 2020
5. EU-project Optemus. http://www.optemus.eu/. 3 Mar 2021
6. Doppler C, Weiß GB, Lorscheider T, Ponchant M, Schönrock P (2020) Predictive thermal-managment in a mild HEV application to reduce energy consumption. In: Automotive thermal management online conference 2020
7. Jakubek S (2020) Predictive energy management of heavy-duty fuel cell trucks fuel cell trucks. In: A3PS eco-mobility 2020
8. A3PS (2018) Eco-mobility 2030 plus—Roadmap
9. Nöst M, Brandstätter B, Eilenberger A, Prenninger P, Trattner A, Wolfbeisser A (2020) A3PS Positionpaper. Spaceplanes, pp 105–123

Michael Nöst is Chairman of IESTA—Institute for Advanced Energy Systems and Transport Applications. In this role he is supporting and coordinating international and national research projects with the research focus on e-mobility, thermal system management and automated driving. As CEO of A3PS—Austrian Association for Advanced Propulsions Systems, he bundles the interests of industry and academia regarding research and development of BEVs, PHEVs, Fuel Cell Vehicles and their respective energy carriers. Michael chairs the international conference Eco-mobility in Vienna for several years and is lecturer at Vienna University of Technology for charging infrastructure.

Christian Doppler holds a master's degree in Urban Technologies with focus on Energy- and Environment systems. His research focus is in the field of Thermodynamics and Thermal-Management for transport systems. His expertise in this fields mainly embraces Thermal-Management strategies for hybrid, electric and fuel cell driven applications. At Virtual Vehicle Research GmbH, Christian is involved in international and national research projects devoted to research on prospective methods to reduce emissions in the transport area.

Alexander Mladek holds a master's degree in Aeronautical Engineering. He is working at Virtual Vehicle Research GmbH as a Senior Researcher in the field of Thermal Management and Mobile Air Conditioning HVAC. His research interests cover efficient thermal management and comfort systems as well as machine learning applications. He is involved in international and national research projects with his special focus on software integration and thermal simulation.

Chapter 16
How Platooning Research Enhances the European Innovation System

Ilja Bäumler, Herbert Kotzab, and Walter Aigner

Abstract Maybe we would need a convincing narrative how innovation and our innovation system contribute to societal wealth—a new kind of Adam Smith equivalent. European road transport is part of a wider ecosystem where significantly increasing levels of digitalisation, automation and innovation will re-shape the world as we have known it. Ambitious political agendas to enhance sustainability and to increase transport effectiveness beyond what can be achieved within a fragmented and traditional way of operation have added momentum. Nevertheless, there are significant open issues beyond what road-maps to various futures maintain to know. Management narratives related to innovation and innovation systems have been challenged. This chapter intends to outline some of the elements how this kind of C-ITS-related platooning research has enhanced our shaping and re-framing of new questions and concepts regarding the European innovation system. Even without electronically coupled trucks on European public roads, elements of dynamic capabilities have evolved. On the other hand, it has become obvious how far some stakeholders have fallen behind the knowledge generation in European C-ITS-related projects. By means of rather selective knowledge intake and knowledge-related search paths, some institutions have shown to be some fifteen years behind accessible knowledge.

Keywords European innovation system · Impact · Knowledge spillover · SAE L2 truck platooning · Truck automation · Logistics

16.1 Introduction

European road transport is part of a wider ecosystem where significantly increasing levels of digitalisation, automation and innovation will re-shape the world as we

I. Bäumler · H. Kotzab
Universität Bremen, Bremen, Germany

W. Aigner (✉)
High Tech Marketing, Vienna, Austria
e-mail: wa@hitec.at

have known it. Ambitious political agendas to enhance sustainability and to increase transport effectiveness beyond what can be achieved within a fragmented and traditional way of operation have added momentum. Nevertheless, there are significant open issues beyond what roadmaps to various futures maintain to know. Management narratives related to innovation and innovation systems have been challenged (see for example [12] or in other contexts [18–20, 37, 38]). Digitalisation of road infrastructure provides one prerequisite for increased effectiveness in a world of connected and assisted mobility (compare also [1, 5, 29, 35]). The European Union has earmarked 20 percent of its post-pandemic recovery funds for digital initiatives. Digitalisation is anticipated to shape the upcoming two decades. In some sense we might feel how far Europe goes in attempting to stimulate innovation. C-ITS—contrary to what some legacy views might suggest—is not one single concept and is seen with rather different expectations and context knowledge in different industries – road operators, traffic management, fleet operation, new mobility and transport service providers, air quality authorities and many more. Some academic knowledge uptake in this field of C-ITS has been rather slow, if compared to project reports in [2–4, 6, 18] and Connecting Austria. Effectively, learning together will continue to be a prime challenge for Europe's innovation system and Europe's economic future. How regions create value will see elements of a new theory and radically opening up of dynamic capabilities in the context of e.g. economic growth. Platooning-related research and innovation will continuously face new questions and white spots in our knowledge. Connecting Austria has demonstrated the potential for bringing in an infrastructure-based view into future forms of road-based freight transport. However, this framing most probably needs to be reinvented in many more ecosystems and knowledge silos. Some examples include (see also our description of contexts in [5, D5.2] and [10]): The inclusion of predictive maintenance in traffic management as well as its ecological role for Europe; changing travel behaviour caused by traffic reduction as a consequence of COVID-19; handling of fluctuating weather conditions or significantly different driver populations due to demographic changes [33].

This chapter intends to outline some of the elements how this kind of C-ITS-related platooning research has enhanced our shaping and reframing of new questions and concepts regarding the European innovation system. Even without electronically coupled trucks on European public roads elements of dynamic capabilities have evolved. On the other hand, it has become obvious how far some stakeholders have fallen behind the knowledge generation in European C-ITS-related projects. By means of rather selective knowledge intake and knowledge-related search paths, some institutions have shown to be some fifteen years behind accessible knowledge. During a recent presentation within the frame of the H2020 ARCADE project Martin Russ from Austria Tech, Vienna has used the term "obfuscation" (creating code that others don't understand") to describe elements of the challenge ahead [11].

Most of this chapter intentionally remains rather implicit and somehow in between the lines. Readers are explicitly invited to use our text as input for taking up and or discussing how we can jointly contribute to making Europe a sustainable ecosystem for innovation, digitalisation and automation.

16.2   Digital Road Infrastructure Leveraging ITS Systems in Europe

16.2.1   Selected Elements of the Current Situation

Some areas in Europe have had an excellent track record in effectively managing the road network as a key element in our context of significantly increased road transport and growing economies. Digitalisation will bring new opportunities for increasing effectivity and efficiency. However, any system at capacity limits will need to mitigate unintended effects from a longer transition period with so called mixed traffic—where most trucks are replaced by newer versions in rather short intervals. Many of these digitally enhanced vehicles with highly professional drivers drive on the same road network with other non-professional road users whose vehicles have rather diverse levels of driver assistance. This allows the mitigation of risks at heavy traffic areas or areas of incidents by making a digitalised road infrastructure in a diverse European traffic landscape possible. However, this requires a high penetration of automated vehicle functions (see [5, 40]). Today, road authorities as well as regional governments can make use of such digital ecosystems making use of such an infrastructure (see our discussions in [5, 33]).

Different dynamics are in place concerning road operators' current situation and key challenges in different regions in Europe. We provided a more extensive picture on the context and current situation as of 2020 in our joint MANTRA WP5 documents. Even though digitalisation can positively affect a country's or even a region's competitiveness, some stakeholders do not focus on this but remain on the level of an efficient and safe road operation. It is however expected that this may change in the near future [5].

16.2.2   Potential Drivers of Socio-technical Transitions Ahead

Digital ecosystems and/or cooperative platforms require digital infrastructures. Mantra [5, D5.2] reports on significant savings of 20 billion € in reduced labour costs, due to less driving hours if real-time traffic avoidance navigation is applied. [24] shows that data economy can make up nearly 90 percent of the EU GDP and important stakeholders are expected to contribute in order to set up a sustainable European data ecosystem (again for more details we refer to our chapter in [5, D5.2]).

Digitalisation is anticipated to contribute to a significantly more heterogeneous vehicle landscape on our roads with various forms of assisted driving and SAE-levels in place even though vehicle manufacturers want to stick on their own services even though there may not be a possibility for cooperation with other stakeholders like networks or different traffic control centres. Vehicle manufacturers plea for upgrading the digital European infrastructure in order to allow automated driving. Consequently, the commitment for investments depend on how first mover behaviour will be rewarded (compare for example [7–9]).

Changing traffic behaviour and actors such as cyclists and/or pedestrians, the increasing importance of ecology as well as demographic changes influence the upcoming of suitable solutions as different types of bottlenecks may occur [5].

All over Europe, freight and passenger traffic is expected to heavily increase and this growth is recognised as a significant driver for an evolving digital infrastructure in the future. Furthermore, changing weather conditions which lead to an increasing number of severe weather events can also be seen as a driver. Besides these, more and more vehicles and service providers apply assisted driving systems based on sensor technology and use more and more AI which will also significantly affect the growth of digital road infrastructures.

At the same time, it is necessary to focus on cyber security issues (see [39]) which will lead to an increased cooperative behaviour between all relevant stakeholders of an ecosystem, even with those that are not primarily a member of a road operator's sphere.

Bishop [13] expects new players outside the existing community of the road transport domain to enter the field. This will also lead to different forms of cooperation. The precise development and design of these new processes and operations depends heavily on who takes the lead in coordinating multiple traffic control systems in a divers European traffic setting [5].

Besides that, it is necessary to consider the different velocity of innovation of different actors like chip manufacturers, mobile network operators, road operators, and other digital service providers when considering the design and operation of digital infrastructures.

External stakeholders are expected to introduce innovative solutions for a digital road infrastructure that can be pooled and thus deployment costs can be reduced. By this, risk can be mitigated and digital networks can be made accessible [5]. This may be a worthfully solution as a full digital infrastructure can not be set up all over Europe due to different traffic densities and road usage.

16.2.3 Particular Demanding Situations for a European Innovation System

The other side of the digitalisation medal refers to the ability of transforming the real world into a digital one as automated decision making is different and require new procedures. While we have studied these phenomena within the CEDR CAD WG in several workshops and documented results in Mantra D5.2 as well as in the final report to the CEDR 2017 call on automation, yet it is uncertain whether existing sensors and automated interactions are capable to deal with the existing complexity or if new types of sensors are required (concept of absorptive capacity; [22]).

It rather looks like a kind of never ending story when it comes the upgrading of a digital infrastructure and a lot of technology representatives see this as a huge challenge, when considering the digitalisation of typical nodes in a value chain such as highways, railways, ports, logistics hubs and/or airports [5].

Another challenge refers to the urban and rural areas of Europe and the acceptance of a full digital infrastructure by their inhabitants. Within the population, we can observe a diminishing acceptance of further improving safety of roads which are considered to be already on a high safety level. The focus could not be only on special interest groups who use high-end vehicles but on all users of a road infrastructure. Such issues have been already heavily discussed e.g. during EUCAD conferences in Brussels 2017 and 2019 [5].

However, national road authorities (NRA) are more and more required to make high quality infrastructures available in order to maintain and to sustain the competitiveness of countries and regions in the future. This shall be realised without waiting until certain technology is available and thus requires first-mover attitudes when it comes to investments. The challenge is even higher, if short innovation cycles are considered [5].

16.2.4   New Roles for Stakeholders

The degree of automated processes and operations depends on how well AI, digitalisation and big data from fixed and mobile sensors are applied. It further depends on the diffusion of such technology within the industry players, which is also inspired by innovative local digital ecosystems. In many cases, existing solutions built upon technology that was acquired some decades ago and was due to safe critical infrastructures as well as less experience of road operators in procuring innovation. Consequently, big tech companies enter more and more the domain of mobility and play an increasing important role in digital mobility ecosystems. One example can be seen in the field of smart cities. NRA's are called to be more active in such areas by being more proactive in their coordination and supervision tasks [5].

16.2.5   Dynamically Evolving Legal Framework

Human decision making has a huge influence on road operators' ITS, as well as on traffic control systems due to the increasing degree of automated processes and operations. The traffic centre processes will be increasingly automated, and by 2030 many traffic management systems are anticipated to being capable of 24/7 operation without any human involvement. However, a legal framework is therefore required [5]. Dynamically evolving customer expectations into home delivery, just-in-time delivery and other newly emerging transport services will further stimulate entrepreneurial ventures. Maybe this is a valid opportunity to look into the landscape of some so-called discrepancies. Will the future show some kind of trade-offs? And what can be eventually kind of re-framed as opportunities?

16.3 Discrepancy Between Customer Requirements and Eco-friendly Transport Logistics

The external conditions of the corporate world have changed considerably in recent years. Digitalisation, globalisation, and fast communication options have led to increasingly fierce competition. Delivery concepts such as just-in-time, drop-shipping or same-day delivery are leading to an increase in transport volumes on the road infrastructure. In addition, the end consumers' increasing request for cosiness favours home delivery, resulting in frequent traffic jams and slow-moving traffic both within and outside the cities [17]. This results in increasingly complex flows of goods, information and merchandise along the value and supply chains. For this reason, closer cooperation between the stakeholders involved is necessary.

However, logistics is also under a certain pressure of expectations from society, since on the one hand a trend towards growing ecological awareness can be observed, but on the other hand not all individuals want to limit themselves. In this respect, from the transport service provider's point of view, the focus lies, particularly on the last mile. Here, an above-average proportion of traffic emissions is produced due to the many starts and stops. Cooperative intelligent transportation systems (C-ITS) make it possible to minimize transport distances and optimise the energy efficiency of trucks [21]. The design and further development of transport systems is becoming an increasingly important task in this context to ensure high quality road transport [14].

Such innovations are necessary in Europe, for example, because of the EU's target, set in 2011, of reducing emissions by 80–95% by 2050 compared to 1990 levels. The overall goal is even to become a net-zero greenhouse emission economy [23]. Since limiting freight mobility is not always considered a valid option, the aim is to increase efficiency through improved and new transport and mobility services based upon C-ITS.

In addition to the expectations of the various stakeholders, the need for intelligent traffic management solutions is also due to the increasing complexity of transport systems, especially in urban areas. First and foremost, it is necessary to ensure an adequate infrastructure to avoid negative effects of transport on the environment [34].

One way to satisfy the rising demand for transportation is to expand the road network. However, for most areas in Europe, this is hardly feasible from an infrastructural and financial perspective [16]. However, as a possible solution to newly emerging capacity bottlenecks, C-ITS gained in importance and is anticipated to significantly shape the current transition period into advanced forms of assisted mobility and transport. The use of C-ITS is intended to make transport logistics safer, more efficient, more flexible, and more sustainable than conventional ways of operating. Another goal is to generate new capacities on all modes of transport [30]. The acronym ITS refers to information and telecommunication technologies as well as control and automation technologies, which, through their combined application with the given infrastructure and the political and legal framework conditions, should contribute to increasing efficiency in the transport sector [26, 27]. Giannopoulos et al. [28] see the potential and goals of ITS primarily in enabling safer and more efficient transport and at the same time reducing emissions. A prerequisite for this is the optimisation of existing supply chains and infrastructure regarding energy-efficient use. This can be supported and further developed by innovative ITS-based tools that control and optimise the planning, organisation, and execution of transports, also from an ecological point of view.

16.3.1   Technical, Legal, and Social Aspects of C-ITS

Technical, legal, or societal requirements (see Table 16.1) will continue to interfere with the process of digitalisation along with the possibilities of intelligent traffic management. The process of digitalisation, which according to experts will be a complete part of our everyday and professional lives by 2030, describes the "increasing data-consistent networking of all areas of the economy, cities, infrastructure and private individuals" [21, p.32]. The so-called Internet of Things, as a result of complete digitisation, refers to the linkage of objects in the transport environment such as goods, vehicles and sensors in devices, so that, for example, precise tracking of shipments or messages on goods status and traffic volumes are possible.

Such networking offers many advantages, especially for logistics service providers, since it shortens delivery times and enables the customer to track the goods handling. The complete digitisation of all objects located in public spaces ultimately culminates in the vision of the Smart City, in which all information on the infrastructure, human and technical actors, as well as events taking place, are summarised in a digital image. Ideally, this would show logistics service providers enormous potential for optimising logistics on the last mile and, if necessary, even enable anticipatory transport logistics. Predictive algorithms that rely on Big Data help logistics service providers to cut delivery times and augment process efficiency and service quality [21, 31].

Table 16.1 summarises the technical, legal, and societal challenges and requirements that need to be considered when deploying C-ITS. For a comprehensive C-ITS implementation, however, it will be necessary to continuously adapt road traffic law to technological progress and at the same time promote acceptance and education in society. In addition, data sovereignty over personal and vehicle related data is of high importance. At any time, road users should be able to know and to control what kind of information they transfer through C-ITS application. Similarly, C-ITS applications must not be misused as a means of surveillance. In the light of cybercrime statistics, C-ITS must meet high security standards.

16.3.2   Critical Discussion of C-ITS and the Needs of Society

First, C-ITS must satisfy several stakeholders due to its central importance, both for inner-city traffic and for long-distance trips outside of cities. Private road users and professional drivers are among the most important representatives, as they are

Table 16.1 Challenges and requirements of C-ITS deployment in the fields of technology, law, and society

Technology	Law	Society
Protection against manipulation	Adaption of road traffic law	Investigation of dilemma situations
International standardisation	Extension of data protection law	Development of public acceptance
Implementation of product liability	Formulation of legal liability regulation	Priority of social interests over individual interests
Further development of intelligent road infrastructure	Control by public authorities	Inclusion of the proper use of automated systems in digital education

directly affected. Irrespective of whether it is a question of goods, service or passenger transport, logistics and software solution providers need to conceptualise C-ITS in such a way that these applications support road users in driving their vehicles, while at the same time apply to safety and security standards. Other stakeholders are city administrations, fleet operators, logistics software solutions providers, road administration, traffic infrastructure operators, terminal and or harbour operators, third party logistics, traffic management software providers and truck manufacturers.

Second, electronic toll collection can also play a role in C-ITS. The introduction of toll roads has two main objectives which are the creating of an income source as well as the control of traffic [15]. The collected toll as a source of income is primarily used for road rehabilitation. However, it is also seen as a fair distribution of costs according to the polluter-pays principle. In addition, tolls can offer alternative routes, separate optional from mandatory traffic, or reduce the overall traffic volume. Regardless of the objective, C-ITS can monitor traffic on corridors or in urban areas and use tolls to uncover hidden road capacities in certain sections of the road and thus regulate traffic. Road administrations could use traffic flow information to charge a fee on specific roads to force usage of driving alternatives.

In terms of emergency driving, C-ITS applications can create a priority for fire departments, rescue and emergency vehicles at traffic lights or intersections. The aim of these systems is to ensure that the rescue services reach their destination early and that subsequent accidents are avoided [36]. In these systems, a device is integrated in rescue and emergency vehicles which interacts with infrared beacons installed along the road. If, for example, emergency vehicles pass such beacons in emergency mode, a signal is sent via the device in the vehicle to the traffic control system so that the signal is changed at the next traffic light [32].

A truck drivers' job comprises many activities such as to control the vehicle, secure the transported goods, load and unload and be aware of their risk potential on streets at any time. Professional truck drivers also perform administrative tasks such as documentation, route planning and customs clearance. Furthermore, they need to adhere to legal regulation on rest and driving times. In rare cases, truck drivers even process order picking in warehouses [25]. Beyond that, the range of tasks is expanded to include communication activities, maintenance, and servicing. With C-ITS, the driver can receive support in his numerous tasks. For example, parking spaces can be reserved for the driver at rest areas depending on his rest and driving periods, in such a way, that he does not have to search for a rest area on his own. It is also conceivable that route guidance through the city, by means of suitable measures such as automatic traffic light phase adjustment, could help to get heavy goods traffic to its destination as quickly as possible. Many application scenarios are conceivable here, in which C-ITS can provide additional support for the driver and optimise traffic management by means of suitable display and guidance systems.

The profitability of a company depends on the efficiency of the executed processes. In addition to logistics efficiency, which provides for an increased utilisation of trucks and the optimisation of the logistics chain, vehicle efficiency, driver efficiency and efficient route planning are other dimensions of efficiency measurement. Vehicle efficiency describes an improvement of the vehicle through improved and new technologies, more effective fuel consumption for increased ranges or a more ergonomic design. In terms of driver efficiency, the driver can be supported by increased training or on-board units that measure driving behaviour. The last form of efficiency describes the optimisation of route planning, where fleet operators plan routes based on traffic density and topography and general data of infrastructure. With C-ITS, all efficiency goals can be achieved through the measures presented.

If all efforts are pulled together, a straightforward realisation remains that widespread deployment of C-ITS is achieved once a critical mass of C-ITS users has been reached. The critical mass is defined here as the required number of C-ITS users, in order for other companies to agree to overcome the investment hurdle associated with the purchase of hardware and software and additional staff training, in order to become part of the overall system. To provide this incentive, pilot projects could be started on busy highway sections. While the deployment of C-ITS can be seen as a genuine attempt to solve the discrepancy between customer demands, sustainability issues and ecological responsibility, it is very important to stress out, that customers need to understand their role in this play. This way the responsibility on mitigating negative effects on traffic environment is shared between all stakeholder, and thus can lead to a joint and satisfying solution.

16.4 Jointly Building Absorptive Capacity in Europe's Innovation System

C-ITS and platooning-related research and innovation will continuously face new questions and white spots in our knowledge. Guiding thread for this chapter has been our question: How does this kind of C-ITS-related platooning research enhance the European innovation system—even without electronically-coupled trucks on the road? There have been significant knowledge spillovers into road operator spheres, digitalisation ecosystems for Europe's future, advanced traffic management, logistics, air quality management. Connecting Austria has contributed to raising awareness with several European key stakeholders on how an ecosystem of innovative organisations in a topographically challenging European road transport hotspot can proactively contribute to a kind of redrawing larger parts of the innovation and research roadmap. Four years ago, it seemed like electronically coupled truck platooning is kind of resolved internationally and European road infrastructure or an ecosystem in a country without truck manufacturers would not have any say. Four years ago the claim was: "It will be on the roads soon." The outlook has significantly shifted. Connecting Austria has demonstrated the potential for bringing in an infrastructure-based view into future forms of road-based freight transport. The bottleneck now looks like how we can effectively manage newly emerging bottlenecks by means of effectively using C-ITS. However, this framing most probably needs to be kind of reinvented in many more ecosystems and knowledge silos.

Dynamically evolving customer expectations into home delivery, just-in-time delivery and other newly emerging transport services together with newly emerging technologies and smart take up we will see interesting decades to come. Digital road infrastructure has the potential to significantly leverage the way we will organise road-based freight transport systems in Europe. But it has the potential for more than that. A European innovation system is made of ecosystems and key stakeholders thinking and conceiving their activities also in terms of joint absorptive capacity. However, the future will be open. Unleash entrepreneurial innovation is anticipated to rank high. And the digital capacity building together with innovation brings the opportunity to attract some of the finest talents in this field of road-based freight transport systems and innovation.

This very chapter as well as this book are an implicit illustration of the opportunities and challenges in learning together in a highly dynamic cross-industry field (C-ITS) where many facets are not fully documented in scientific literature. This process of integrating views from one selected team of our distinguished board of scientific advisors in Connecting Austria will be continued by integrating views from our various audiences in the months to come.

References

1. DIGEST (2021) Digitaler Zwilling des Verkehrssystems Straße. https://projekte.ffg.at/projekt/3894859. 10 Feb 2021
2. Elektronische Deichsel–Digitale Innovation–EDDI. https://www.bmvi.de/SharedDocs/DE/Artikel/DG/AVF-projekte/eddi.html. 10 Jan 2021
3. ERTICO: Multi-brand Truck Platooning to become reality in Europe. https://erticonetwork.com/multi-brand-truck-platooning-become-reality-europe-eu-co-funded-ensemble-project/
4. HelmUK, a platooning project under the technical leadership of Highways England: Truck Platooning: the future of road transport. https://www.smmt.co.uk/2020/06/has-truck-platooning-hit-the-end-of-the-road/
5. MANTRA – making full use of automation for national road transport authorities: 2020. https://www.mantra-research.eu/

6. Sweden4platooning 2020 public report. https://www.saferresearch.com/index.php/library/sweden4platooning-public-report
7. ACEA. 25 actions for a successful restart of the eu's automotive sector. https://www.acea.auto/publication/paper-25-actions-for-a-successful-restart-of-the-eus-automotive-sector/
8. ACEA (2019) Automated driving. roadmap for the deployment of automated driving in the european union. https://www.acea.be/publications/article/roadmap-for-the-deployment-of-automated-driving-in-the-european-union
9. ACEA. Roads of the future for automated driving. ACEA Discussion paper 2020. [currently not available online https://www.acea.auto/nav/?content=publications [an archive copy is available from the authors of this chapter]
10. Aigner W, Kulmala R, Ulrich S (2019) Vehicle fleet penetrations and odd coverage of NRA-relevant automation functions up to 2040. mantra: Making full use of automation for national transport and road authorities—NRA core business, deliverable 2.1
11. ARCADE. Workshop on common evaluation methodology for automated driving tests. esp. Martin Russ: Obfuscation: writing code that nobody understands. Recording available at https://www.youtube.com/watch?v=QtHvYvkrsfc
12. Berkun S (2007) The myths of innovation. O'Reilly Media; Sebastopol, CA
13. Bishop R (2020) Automated driving decades of research and development leading to today's commercial systems. In: Fisher DL, Horrey WJ, Lee JD, Regan MA (eds) Handbook of human factors for automated, connected, and intelligent vehicles. CRC Press
14. Boltze M, Schäfer PK, Wolfermann A (2006) Leitfaden Verkehrstelematik. Hinweise zur Planung und Nutzung in Kommunen und Kreisen. Internationales Verkehrswesen, 58 (10), S. 465–468. DVV Media Group
15. Boursas L (2009) Trust-based access control in federated environments. Technische Universität München
16. Bäumler I, Kotzab H (2020) Intelligent transportation system development, main influences and key technologies. In: Nofoma 2017 the 29th Nofoma conference: taking on grand challenges. Lund University, pp 830
17. Bäumler I, Kotzab H (2020) Scenario-based development of intelligent transportation systems for road freight transport in Germany. In: Urban freight transportation systems. Elsevier
18. C-ROADS (2019) C-roads—the platform of harmonised c-its deployment in Europe. https://www.c-roads.eu/platform.html
19. Canton J (2015) Future smart. Managing game-changing trends that will transform your world. Da Capo Press
20. CEDR (2017) Description of research needs (dorn). cedr transnational road research programme call 2017: Automation. cedr transnational road research programme funded by austria. and the United Kingdom
21. Clausen U, Stütz S, Bernsmann A, Heinrichmeyer H (2016) ZF-Zukunftsstudie 2016 Die letzte Meile. ZF Friedrichshafen AG. Friedrichshafen
22. Cohen WM, Levinthal DA (1990) Absorptive capacity: a new perspective on learning and innovation. In: Administrative science quarterly, vol, 35, issue no. 1, pp 128–152 (special issue: technology, organizations, and innovation
23. European Commission (2018) Communication from the commission to the European parliament, the European council, the council, the European economic and social committee, the committee of the regions and the European investment bank-a clean planet for all a European strategic long-term vision for a prosperous, modern, competitive and climate neutral economy
24. European Commission (2020) A European strategy for data. communication from the Commission to the European Parliament, the Council, the European Economic and Social Committee and the Committee of the Regions. com (2020) 66 final. available online at: https://op.europa.eu/en/publication-detail/-/publication/ac9cd214-53c6-11ea-aece-01aa75ed71a1/language-en
25. Flämig H (2015) Autonome Fahrzeuge und autonomes Fahren im Bereich des Gütertransportes. In: Maurer M, Gerdes JC, Lenz B, Winner H (eds) Autonomes Fahren. Springer, Berlin, pp 377–398

26. Gattuso D, Pellicanó DS (2014) Advanced methodological researches concerning its in freight transport. Procedia—social and behavioral science, transportation: can we do more with less resources? In: 16th meeting of the Euro working group on transportation, vol 111, pp 994–1003 (Porto)
27. Giannopoulos GA (2009) European its for freight transport and logistics: results of current eu funded research and prospects for the future. Eur Transp Res Rev 1:147–161
28. Giannopoulos G, Mitsakis E, Salanova Grau JM (2012) Overview of intelligent transport systems (ITS) developments in and across transport modes. JRC Scientific and Policy Reports
29. Hjälmdahl M, Svedlund J, Mattsson F, Bahr M, Baid V (2020) In: EU EIP Activity 4: harmonisation ClusterSub-activity 4.2: facilitating automated driving summary of impacts, benefits and costs of highly automated driving
30. Janker CG, Lasch R (2001) Telematik im Straßengüterverkehr—stand der Umsetzung und Nutzenpotenziale. In: Sebastian H-J, Grünert T (eds) Logistik management. Vieweg+Teubner Verlag, Wiesbaden, pp 245–255
31. Heutger M, Kückelhaus M, Zeiler K, Niezgoda D, Chung G (2014) Self-driving Vehicles in Logistics—A DHL Perspective on Implications and Use-cases for the Logistics Industry. DHL Trend Research, Troisdorf, Germany
32. Kotani J, Yamazaki K, Jinno M (2011) Expanding fast emergency vehicle preemption system in Tokyo. In: 18th ITS World CongressTransCoreITS AmericaERTICO-ITS EuropeITS Asia-Pacific
33. Kulmala R, Ulrich S, Penttinen M, Rämä P, Aigner W, Carsten O, Tuin M, Farah H, Appel K (2020) D5.2 road map for developing road operator core business utilising connectivity and automation project nr. In: 867448 MANTRA: Making full use of Automation for National Transport and Road Authorities—NRA core business
34. Małecki K, Iwan S, Kijewska K (2014) Influence of intelligent transportation systems on reduction of the environmental negative impact of urban freight transport based on Szczecin example. Procedia—social behavioural science, vol 151. Green Cities—Green Logistics for Greener Cities, Szczecin
35. Malone K, Schreuder M, Berkers F, Helfert K, Radics L, Boehm M (2019) Digitalisation and automation. implications for use cases. Identification of Stakeholders and Data Needs and Requirements. DIRIZON Deliverable (3.1. Draft 0.7)
36. Moerke A (2007) Intelligent transport system its. In: Moerke A, Walke A (eds) Japans Zukunftsindustrien. Springer, Berlin, Heidelberg, pp 275–286
37. CCAM Partnership (2020) Strategic research and innovation agenda. proposed European partnership under horizon Europe CCAM connected. Coop Autom Mobilty 9
38. Sainsbury D (2020) Windows of opportunity. How nations crate wealth. Profile Books
39. Strand S, Zofka E, Ponweiser W, Lamb M, Hedhli A, Adesiyun A (2020) Practical learnings from test sites and impact assessments. Technical Report 4
40. Ulrich S, Kulmala R, Aigner W, Wegscheider S (2017) Final programme report CEDR transnational road research programme call 2017 automation MANTRA. DIRIZON. STAPLE. CEDR Contractor Report

Ilja Bäumler is Postdoctoral Researcher at the University of Bremen. His research focuses on Intelligent Transportation Systems (ITS) and Supply Chain Management. In the research area of ITS he is particularly interested in possible future scenarios and their implications on stakeholders. His dissertation "Development paths of intelligent transport systems for road freight transport" was published at the State and University Library Bremen in 2019.

Herbert Kotzab is a Professor at the Chair of Logistics Management at the University of Bremen and International Professor at Othman Yeop Abdullah Graduate School of Business, Universiti Utara Malaysia, Malaysia. He received his Master of Business Administration in Marketing and Management, PhD (1996) and Postdoctoral degree (Habilitation: 2002) from the Vienna University of Economics and Business Administration. Prior to his assignment at Bremen, he held a Professor position at Copenhagen Business School at the Department of Operations Management. His research focuses on Supply Chain Management, Service Operations and Consumer Driven Value Networks. Since 2013, he has been a member of the Editor-in-Chief-Board of Logistics Research.

Walter Aigner has been a kind of boundary-spanning individual [between pioneering users, public administration, research and various industries] since the early 1990ies. As managing director at HiTec he prepared several national and European innovation and technology programmes and serves as an independent expert on evaluation and impact assessment. His focus is on key individuals in the European innovation system and how they nourish our commitment to contributing to a more nuanced answer for Europe's share in a global challenge to effectively cooperate with the US and Asia in a highly competitive environment of innovation, digitalisation and automation.

Chapter 17
Discussion

Walter Aigner, Matthias Neubauer, and Wolfgang Schildorfer

Abstract In 2017, the Connecting Austria project was internationally unique with respect to the special consideration of the infrastructure and traffic perspective as well as the special consideration of investigating an urban truck platooning use case with traffic-light-controlled intersections before and after motorway entrances. The three main target groups of the project were: (1) road operators/infrastructure providers, (2) logistics operators and (3) C-ITS industry. Especially for those target groups and policy maker faced one central question at that point in time —"How can safe truck platooning reduce CO_2-emissions and how can this help to strengthen the stakeholders' role in their market or political environment?". Cooperative, connected and automated mobility shape the future of road transport. Thereby, truck platooning represents an important application case in the transport logistics domain. In this chapter, the research and evaluation results presented in this book are discussed along the following three fundamental pillars: (1) traffic safety and legal issues, (2) sustainability and (3) truck platooning deployment. Finally, limitations and cultural blind spots experienced within international workshops and discussions in the context of the Connecting Austria project are reflected.

Keywords Traffic safety · Legislation · Sustainability · Truck platooning deployment

17.1 Traffic Safety and Legal Issues

Especially when introducing new technologies in the domain of automated driving, a close observation of a range of parameters is desirable—and even mandatory. In the case of truck platooning, this is not only referring to the automated system, but

W. Aigner
High Tech Marketing, Lothringerstraße 14/6, 1030 Vienna, Austria

M. Neubauer · W. Schildorfer (✉)
Department of Logistics, University of Applied Sciences Upper Austria, Steyr, Austria
e-mail: wolfgang.schildorfer@fh-steyr.at

© The Author(s) 2022
A. Schirrer et al. (eds.), *Energy-Efficient and Semi-automated Truck Platooning*,
Lecture Notes in Intelligent Transportation and Infrastructure,
https://doi.org/10.1007/978-3-030-88682-0_17

also to the temporarily inactive driver and the interaction with other road users. The impact on driver attention and effects of risk compensation or adverse behavioural changes of other road users in reaction to platoons, etc., cannot be ruled out yet. Either way, systematic collection of a wide range of safety-relevant indicators within a cross-border longitudinal field test will be key for truck platooning and the safety of all road users involved. Quite a lot of safety-related material have been collected in various European research projects and even industry-driven studies in the USA. Overall, truck platooning seems to be an enabler for increasing traffic safety on public roads. However, in advance, a major issue needs to be solved—the harmonisation of European law with regard to truck platooning, e.g. for regulating minimum distances between trucks, labour law or technical preconditions for building a platoon.

Although road safety might not be the primary motivation for proponents of platooning, it is worth discussing potential safety benefits. Human errors are a leading cause of road accidents; e.g. in Austria, inattention and distraction alone accounted for 51% of HGV crashes on Austrian motor and expressways between 2014 and 2018. Thus, relevant potentials to reduce truck accidents exist. However, one unknown factor in the equation is the reliability of systems that control the connected/automated driving function as well as the assumed penetration rate.

In order to enable safe, high-performance and efficient truck platooning control concepts, the global properties (surrounding traffic, infrastructure, platoon dynamics, road properties and route) must be appropriately considered in the planning and optimisation of platoon trajectories. For the effective realisation of these movement patterns in vehicle control, the distributed or locally acting control on the individual vehicle level must be combined with the essential information from the broader, global context in a suitably prepared form. Cooperative platoon control strategies make use of provided information from vehicle-to-everything (V2X) communication to reduce energy or fuel consumption, increase traffic flow and improve traffic safety. Thereby, local information and predictions can be shared with the entire platoon, thus improving the effectiveness of the distributed control actions.

Within Connecting Austria, a distributed control concept for cooperative platooning was developed that combines trajectory optimisation and local model-predictive control of each vehicle. The presented control architecture ensures collision safety by design, platoon efficiency and situational awareness with the option of exploiting V2X communication. The resulting platoon control performance was tested and validated in a realistic setting by utilising a co-simulation-based validation framework with detailed vehicle dynamics.

In the Connecting Austria project, the urban truck platooning use case considering intersections has been the most complex one with respect to possible C-ITS, traffic and vehicle control actions. For this use case, platoon control concepts and innovative means to monitor and assess real-time traffic have been developed. Recent progress in video-based vehicle sensors allow for a detailed observation of road users on intersections in urban areas. By combining the measured real-life traffic situation with thorough traffic simulations, a cooperative system design for the dynamic management of traffic flow including vehicle platoons is possible. A video-based traffic flow estimation system was developed and tested at a three-way intersection in the small city of Hallein, Austria. The installed system is able to collect comprehensive information about the traffic situation in near real time. This information can be used to estimate traffic density as well as traffic flows of cars and trucks with high precision. Furthermore, it allows inform cooperative platooning control strategies and support situational awareness.

The comprehensive scenario-based approach taken in Connecting Austria allowed for an effective and efficient development as well as the validation of complex, cooperative control functions in connected and automated driving. The conducted studies do not yield a single result, but instead depend on many parameters (such as platoon spacing/gap policy, surrounding traffic density and speed and many more) and are investigated in terms of the results' sensitivities on these parameters. This approach allows to draw meaningful conclusions despite the inherent uncertainty and spread of the influencing parameters. By using representative conditions, the resulting key performance indicator distributions may be evaluated and interpreted. Under certain circumstances, platooning may lead to extra risks with respect to safety. One such issue are vehicle cut-ins, e.g. in case of a highway exit. For passenger cars, which have to exit the highway, it would be beneficial not to overtake a platoon just before the aimed exit, in order to avoid wriggling through narrowly driving trucks of the platoon. However, what precisely does "just before" mean? The answer depends on the length of the platoon, the speed difference, the deceleration capabilities of the cars, the cooperative behaviour of the single trucks, etc. For such special situations, characteristic diagrams are suitable and have been developed. They allow the estimation of necessary distances before an exit and help specifying adequate distances before special locations and conditions, respectively. Additionally, they are useful in case of danger zones, which demand for the dissolution of platoons or for temporal/spatial increase/decrease of intra-platoon distances.

A further question within the platooning discourse is the ideal distance between the trucks of a platoon. Alongside economic and environmental considerations, this is an important road safety question. Not only in terms of potential malfunction of the used technical equipment but also for the perception of car drivers who for example have to decide at which gap sizes between trucks, they would merge onto the rightmost lane for the purpose of exiting the motorway or merge into traffic coming from an on-ramp. To gain knowledge about the drivers' decision making regarding the acceptance of different gap sizes between trucks, an on-road driving study was conducted in summer 2020. However, in this contribution, only a basic outline of the study design was given, since the data analyses were still ongoing.

17.2 Sustainability

Sustainability is one of the main topics discussed globally right now—and it is necessary to do so. Typically, sustainability is discussed along the pillars (1) economic-, (2) social- and environmental sustainability. Truck platooning may contribute to each of these pillars. The results from Connecting Austria and other research projects confirm that truck platooning allows reducing fuel consumption. As such, truck platooning may contribute to reduce fuel costs (cf. economic sustainability) and CO_2 emissions (cf. environmental sustainability). To this regard, the book presented a fuel assessment methodology and its application. Thereby, the methodological approach covered three main aspects: (1) the assessment of the road infrastructure in terms of the suitability of road segments for truck platooning, (2) the assessment of driving behaviour and strategies for truck platoon formation and dissolution and (3) the assessment of efficiency in terms fuel savings for certain routes. Within an initial method application, a route analysis for an Austrian fleet operator was performed including the assessment of feasible and economic viable routes and scenarios. Furthermore, potential fuel consumption savings and CO_2 emission savings were discussed within the given case.

When analysing a typical route (Origin: Pasching— Destination: Guntramsdorf) of an Austrian fleet operator with respect to the feasibility of platooning and the economic viability of platooning segments, the results indicate a wide range of feasible segments (from 65% up to 92%). The highest share of economic viable road segments is gained when applying a medium formation and dissolution strategy for truck platoons. However, dynamic C-ITS-based truck platoon regulations may even more increase the savings. Overall, the saving potential may be increased via dynamic C-ITS-based truck platoon regulations, instead of statically defined, too restrictive regulations.

Subsequent to the assessment of the feasibility and economic viability of road segments, an analysis of potential costs and emission savings was conducted. This analysis confirmed that in most instances the medium formation/dissolution strategy for a 3 truck platoon driving at an intra-platoon distance between 1 and 1.5 s at a speed of 80km/h is suitable. For the route "Pasching -> Guntramsdorf", the maximum achievable fuel saving is 4.83% for all three trucks in a platoon driving from Linz to Vienna. The minimum saving when applying the fast formation/dissolution strategy at an intra-platoon distance of 1 s leads to fuel savings of 2.53%. The analysis of the fuel reduction also provides a basis for assessing potential savings in CO_2 emissions. Based on the consumption values and a factor for translating fuel consumption in emissions, one may illustrate potential CO_2 savings. Maximum potential CO_2 savings for a three-truck platoon are up to 25kg CO_2 savings (Well-To-Wheel).

The given fuel efficiency assessment results rely on a fuel reduction model for truck platoons. With regard to related work, the range of fuel savings reported is quite high for average savings in a platoon of three trucks at a distance of 15 m (about 2% in the Companion project up to 11% measured within Japan ITS Energy 2016). In the Connecting Austria project, a fuel reduction model for truck platoons has been developed taking into account related work, CFD simulations as well as the validation of measures on a test track.

In addition to economic and environmental sustainability aspects, truck platooning may also affect social sustainability. In this regard, the book summarised related work with respect to technology acceptance of truck platooning. Related work indicated that truck platooning could decrease workload for truck drivers and support safe driving. However, adequate user-centred implementation processes within organisations and awareness building will be crucial for successfully deploying truck platooning.

17.3 Truck Platooning Deployment

What are the next truck platooning-related deployment steps internationally and especially in Europe for the next years? Some years ago, automotive industry pushed the topic with European initiatives and projects like the truck platoon challenge or the up-and-running project ENSEMBLE. Currently, press releases of global automotive industry player pushes more higher automation levels of trucks on closed areas. In parallel, start-ups in the USA show first implementation of truck platooning over long distances on highways. The reasons for this more hesitant attitude of some industry players in Europe could be based on missing legal conditions to maintain platooning in Europe. This ends up in a non-existing market need and seen from a business-perspective a logic strategic change. Some automotive industry players argued that potential savings of platoons are not relevant for customers and therefore officially decided to stop working on that. However, as we claimed above. It is not just a matter of energy savings and cost savings for the trucks. It is a matter of future necessity to reduce emissions to survive the international competition in logistics. Logistics operators need to increase drivers' safety, to even increase the attractiveness of truck driving and finally yet importantly, to reduce costs based on avoided crashes.

Besides the discussion from an industry perspective outlined in the previous paragraph, some more questions need to be addressed when talking about deployment. Taking a business-related perspective and mirroring the typical Austrian and even European situation of quite a lot of small medium enterprises in the logistics domain, the following questions arises. How should they implement new technology in their traditional logistics provider's processes—especially when not yet implemented digital processes or just maintaining a small number of trucks? This question is closely interconnected with the not yet defined truck platooning service business model. Right now, there are some ideas on how to organise truck platoons ad hoc on the road with different brands and companies. However, how long does it take to set-up a player to takeover that service provider role is yet unclear.

With ASFINAG's roll-out of C-ITS stations at 2,000 highway kilometres in November 2020, the C-ITS deployment started in Austria, and the foundation for Day-1 and Day-2 C-ITS services like cooperative platooning was established. Even if the initial investment in Austria took already place, viable business models for the operation, maintenance, and the enhancement of C-ITS services are still an open issue. Examples in existing publications propose as revenue streams taxes/tolls, reduced road maintenance costs or earnings from data provision services. However, when developing/deploying new C-ITS services, viable business models should be investigated to support road operators and technology providers in taking informed decisions and sustainably deploying C-ITS services.

C-ITS and platooning-related research and innovation will continuously face new questions and white spots in our knowledge. There has been significant knowledge spillover into road operator spheres, digitalisation ecosystems for Europe's future, advanced traffic management, logistics and air quality management.

Dynamically evolving customer expectations into home delivery, just-in-time delivery and other newly emerging transport services together with newly emerging technologies and smart take up we will see interesting decades to come. Digital road infrastructure has the potential to significantly leverage the way we will organise road-based freight transport systems in Europe. However, it has the potential for more than that. A European innovation system is made of ecosystems and key stakeholders thinking and conceiving their activities also in terms of joint absorptive capacity. However, the future will be open. Unleash entrepreneurial innovation is anticipated to rank high. Moreover, the digital capacity building together with innovation brings the opportunity to attract some of the finest talents in this field of road-based freight transport systems and innovation.

17.4 Some Limitations and Cultural Blind Spots

When you prepare a large-scale flagship project, probably all key players accept the challenge to contribute to a better world and a better future for Europe. This involves also citizens living in Europe and more specifically in a kind of transit country with challenging alpine topography—like Austria. However, our concepts of what a better world and future would look like have also evolved rather dynamically.

Looking back, we see significant progress as well as entirely new challenges and questions. Here, we will reflect on some of the results towards this overall challenge as well as on selected limitations and blind spots. One guiding frame can be seen in any innovation, or any newness is kind of an intermediate transition. Counterintuitively, this transition can be validly framed rather not a transition between two stable situations, but an ongoing transition in itself.

Connecting Austria was designed to cooperate with all existing activities and to try to cross-fertilise towards specific aspects of a transit country with innovative ITS infrastructure to mitigate unintended consequences from platooning technologies. Coming to some limitations and blind spots in the public discourse on truck platooning and the future of truck platooning: several European stakeholders wonder why the USA is rather readily implementing truck platooning at rather high average truck speeds, when far higher gains in fuel consumption could be made by reducing the maximum speed to European standard levels of 80 km/h. For stakeholders in the USA, this is not even considered a valid proposal. The superficial diagnosis would look into lack of climate-related ambitious commitments in the USA. A more elaborated diagnosis would look into road-freight systems that have been kind of optimised for local bottlenecks. At lower speeds, truck drivers would not make it back home—something that is less obvious for truck operations in Europe; and with increasing transport volumes, hardly anybody would like to increase the number of trucks to make lower speeds feasible under a climate-related agenda. One kind of elephant in the discussion room seems to be climate-related impacts: we often hear 8 per cent or 15 per cent or 35 per cent improvement is not enough. With increasing volumes of goods transported, it is not efficiency but rather effectiveness—across multiple transport modes. Several initiatives have looked into feasibility of modal shift in European freight transport (e.g. ALICE) and have quickly reached practical limitations and cultural bottlenecks. A study group with a Swiss rail operator estimates efficiency gains in the road-based freight transport sector in the order of 35 per cent until 2030. This is anticipated to foster innovation in the rail sector—otherwise it would significantly loose share in the modal split. Requests to limit innovation in the road-based freight sector remain.

Some countries have defined their future-proof road network as part of their innovation system or at least as enabler for sustainable future wealth creation. Stakeholders in Finland currently are studying how a roll out of longer truck platoons will impact infrastructure design at borders, harbours, highway intersections and ramps and entire resting areas and parking lots. This can be partly attributed to a strong forestry industry and a strategic importance of 24-hour global delivery requirements for Scandinavian fish shipped to customers in Japan and South East Asia. As a consequence, in Finland, several stakeholders are preparing for higher levels of truck platooning—even without commercially available offers on the market yet for even lower SAE-Levels in Europe. Europe-wide, several leading fleet operators and logistic operators have already reached higher levels of automation within their logistic hubs. They have expressed that they would immediately adopt automated hub-to-hub truck transport including truck platooning as soon as legal regulations would make this feasible.

In other countries, automation has been perceived as unnecessary risk or detrimental for social inclusion: some stakeholders have explicitly expressed they would define any legal requests into innovative forms of road-based traffic, in a way to kind of prevent the commercially valid roll-out of any automated driving technology—be it in trucks or individual transport. Another elephant in the discussion room is the actual following distance between trucks on European highways. Several anecdotal evidence stories in large printed media have indicated that the actual following distance in 2017 has been significantly lower than what has been demonstrated in electronically coupled high-tech trucks within EDDI in Germany, Drive-Sweden or in the USA. Significant cohorts of truck drivers have a rather high work ethic and maintain this professional mind-set when using assistive systems (validated and confirmed in EDDI). On the other hand, during the 2020 COVID-19 impact phase, driver availability has dynamically evolved. With longer waiting times at national and even regional borders and significant shortages with the most experienced drivers, there is an increasing share of less experienced truck drivers adding momentum. In combination with some of the other mentioned elephants, the value perception of assistive systems and electronically coupled trucks might quickly see significant changes in public discourse.

Digitalisation and steps towards using digital twins for entirely new forms of decision making at road operators have the potential to open up the opportunity space for simulation-based quick responses to the way road-based freight transport is accepted on highways. The CEDR working group on CAD as well as C-ITS groups at ASFI-NAG have proactively addressed several of these opportunities. Results from Connecting Austria were taken up within CEDR's automation study project MANTRA. Three road authorities in Germany, Switzerland and Austria have committed to a detailed study into integrated digital twins for purposes of traffic management and especially ODD-aware forms of traffic management or complex traffic scenarios.

The Connecting Austria team's knowledge generation would not have been possible without the support from several workshops with Richard Bishop and some fellow truck platooning projects (EDDI; Sweden4Platooning). Michael Nikowitz has successfully supported the flagship project within the dynamically evolving complexities in different ministerial High Administrations since 2017. Connecting Austria has opened European awareness on infrastructure-based views and on specific needs for ambitious climate agendas in topographically challenging regions. Several projects now have taken up validation exercises and the concept of infrastructure-based information services to accompany the next wave of digitalisation, automation and future mobility. And, it is steadily evolving how this nourishes our commitment to contributing to a more nuanced answer for Europe's share in a global challenge to effectively cooperate with the USA and Asia in a highly competitive environment of innovation.

Walter Aigner has been a kind of boundary-spanning individual [between pioneering users, public administration, research and various industries] since the early 1990ies. As managing director at HiTec he prepared several national and European innovation and technology programmes and serves as an independent expert on evaluation and impact assessment. His focus is on key individuals in the European innovation system and how they nourish our commitment to contributing to a more nuanced answer for Europe's share in a global challenge to effectively cooperate with the US and Asia in a highly competitive environment of innovation, digitalisation and automation.

Matthias Neubauer is a professor at the University of Applied Sciences Upper Austria for logistics information systems. His research interests cover human-computer interaction, intelligent transportation systems as well as cooperative, connected and automated mobility. Matthias is involved in international and national research projects and teaches in the master's program digital transport and logistics management classes such as process management, distributed logistics systems or geoinformation systems.

Wolfgang Schildorfer is a person who likes the road he still walks to find new chances. Since October 2018, he has been Professor for Transport Logistics and Mobility at the University of Applied Sciences Upper Austria. His research focus is on innovation, business models and evaluation in transport logistics, smart hyperconnected logistics systems, (urban) mobility, sustainable transport systems and new technology markets (C-ITS, CCAM, automated driving, truck platooning).

Correction to: Energy-Efficient and Semi-automated Truck Platooning

Alexander Schirrer⊙, Alexander L. Gratzer⊙, Sebastian Thormann⊙, Stefan Jakubek, Matthias Neubauer⊙, and Wolfgang Schildorfer⊙

Correction to:
A. Schirrer et al. (eds.), *Energy-Efficient and Semi-automated Truck Platooning*, **Lecture Notes in Intelligent Transportation and Infrastructure,**
https://doi.org/10.1007/978-3-030-88682-0

In the original version of the Chapter 5 (Truck Platoon Slipstream Effects Assessment) and Chapter 6 (Validation of Truck Platoon Slipstream Effects), the author "Dr. Christoph Irrenfried" name was not included as a co-author. This has now been rectified and the author's name has been included.

The chapters and the book have been updated with the changes.

The updated original versions of these chapters can be found at
https://doi.org/10.1007/978-3-030-88682-0_5
https://doi.org/10.1007/978-3-030-88682-0_6

Printed in the United States
by Baker & Taylor Publisher Services